Springer Theses

Recognizing Outstanding Ph.D. Research

For further volumes:
http://www.springer.com/series/8790

Aims and Scope

The series "Springer Theses" brings together a selection of the very best Ph.D. theses from around the world and across the physical sciences. Nominated and endorsed by two recognized specialists, each published volume has been selected for its scientific excellence and the high impact of its contents for the pertinent field of research. For greater accessibility to non-specialists, the published versions include an extended introduction, as well as a foreword by the student's supervisor explaining the special relevance of the work for the field. As a whole, the series will provide a valuable resource both for newcomers to the research fields described, and for other scientists seeking detailed background information on special questions. Finally, it provides an accredited documentation of the valuable contributions made by today's younger generation of scientists.

Theses are accepted into the series by invited nomination only and must fulfill all of the following criteria

- They must be written in good English.
- The topic should fall within the confines of Chemistry, Physics, Earth Sciences, Engineering and related interdisciplinary fields such as Materials, Nanoscience, Chemical Engineering, Complex Systems and Biophysics.
- The work reported in the thesis must represent a significant scientific advance.
- If the thesis includes previously published material, permission to reproduce this must be gained from the respective copyright holder.
- They must have been examined and passed during the 12 months prior to nomination.
- Each thesis should include a foreword by the supervisor outlining the significance of its content.
- The theses should have a clearly defined structure including an introduction accessible to scientists not expert in that particular field.

Max García Melchor

A Theoretical Study of Pd-Catalyzed C–C Cross-Coupling Reactions

Doctoral Thesis accepted by
the Autonomous University of Barcelona, Spain

Springer

Author
Dr. Max García Melchor
Department of Chemistry
Autonomous University of Barcelona
Edifici Cn, 08193 Cerdanyola del Vallès
Spain

Supervisors
Prof. Agustí Lledós
Prof. Gregori Ujaque
Department of Chemistry
Autonomous University of Barcelona
Edifici Cn, 08193 Cerdanyola del Vallès
Spain

ISSN 2190-5053 ISSN 2190-5061 (electronic)
ISBN 978-3-319-01489-0 ISBN 978-3-319-01490-6 (eBook)
DOI 10.1007/978-3-319-01490-6
Springer Cham Heidelberg New York Dordrecht London

Library of Congress Control Number: 2013944549

Printed on acid-free paper

Springer is part of Springer Science+Business Media (www.springer.com)

Parts of this thesis have been published in the following journal articles:

1) B. Fuentes, **M. García-Melchor**, J. A. Casares, G. Ujaque, A. Lledós, F. Maseras, P. Espinet. "Palladium Roundtrip in the Negishi Coupling of *trans*-[PdMeCl($PMePh_2$)$_2$] with ZnMeCl: Experimental and DFT Study of the Transmetalation Step". *Chem. Eur. J.* **2010**, *16*, 8596–8599.

2) **M. García-Melchor**, G. Ujaque, F. Maseras, A. Lledós. In "Phosphorus Compounds: Advanced Tools in Catalysis and Material Sciences". Series: Catalysis by Metal Complexes (CMCO), M. Peruzzini, L. Gonsalvi, Eds., Springer-Verlag (Berlin): **2011**, Vol. *37*, Chapter 3, pp 57–84.

3) **M. García-Melchor**, B. Fuentes, A. Lledós, J. A. Casares, G. Ujaque, P. Espinet. "Cationic intermediates in the Pd-Catalyzed Negishi Coupling. Kinetic and DFT Study of Alternative Transmetalation pathways in the Me-Me Coupling of $ZnMe_2$ and *trans*-[PdMeCl($PMePh_2$)$_2$]". *J. Am. Chem. Soc.* **2011**, *133*, 13519–13526.

4) **M. García-Melchor**, M. C. Pacheco, C. Nájera, G. Ujaque, A. Lledós. "Mechanistic Exploration of the Pd-Catalyzed Copper-Free Sonogashira Reaction". *ACS Catal.* **2012**, *2*, 135–144.

5) **M. García-Melchor**, X. Solans-Monfort, G. Ujaque. In Comprehensive Coordination Chemistry-II, J. Reedijk, K. Poeppelmeier, Eds., Elsevier: **2013**, Vol. *9*, Chapter 9.32. *In press*.

6) **M. García-Melchor**, A. A. C. Braga, A. Lledós, G. Ujaque, F. Maseras. "Computational Perspective on Pd-Catalyzed C-C Cross-Coupling Reaction Mechanisms". *Acc. Chem. Res.* **2013**. doi:10.1021/ar400080r.

Dedicated to Rosa, Emma
and the rest of my family

Supervisors' Foreword

Pd-catalyzed C–C cross-coupling reactions have become one of the most useful tools for organic synthesis. Their development was recognized with the Nobel Prize in Chemistry in 2010 awarded to Profs. R. F. Heck, E. Negishi, and A. Suzuki. Applications of these reactions are found in several fields of social interest as agrochemicals, pharmaceuticals, etc. These reactions can be classified depending on the metal or semi-metal present in the nucleophile. For instance, Stille reaction is tin-mediated, Negishi zinc-mediated, etc. The reaction mechanism is generally described in three steps: oxidative addition, transmetalation, and reductive elimination. A deeper understanding of reaction mechanisms is mandatory to control and improve chemical processes. In the present Ph.D. dissertation theoretical calculations have been used to determine, elucidate, and propose mechanisms for Pd-catalyzed C–C cross-coupling reactions. Reaction intermediates and transition states involved in the Negishi, the copper-free Sonogashira, and the asymmetric version of the Suzuki coupling have been characterized, providing a detailed picture of their reaction mechanisms. These results can guide further experimental investigations.

Cerdanyola del Vallès, May 2013

Prof. Agustí Lledós
Prof. Gregori Ujaque

Acknowledgments

No man succeeds without a good woman behind him.
Wife or mother, if it is both, he is twice blessed indeed.

Harold MacMillan

First of all, I would like to thank all those who have supported me, in one or another way, during the course of this Ph.D. thesis and who have made it possible. In particular, I would like to thank my Ph.D. supervisors, Profs. Agustí Lledós and Gregori Ujaque, for introducing me to the exciting world of Research and their support and guidance during this period. I also sincerely thank all my friends from the Chemistry Department at the Universitat Autònoma de Barcelona, especially Carles, Aleix, Sergi, Salva, Laia, Manu, Eli, Víctor, Rosa, Fede, Iker, Jacob and Fran, and my friends from outside the department, Mª José, Manel, Fer, José, and Jessy.

I am also very grateful to all our experimental collaborators for their contribution to this thesis, notably Profs. Pablo Espinet and Juan A. Casares from the Universidad de Valladolid, Prof. Carmen Nájera from the Universidad de Alicante, and Profs. Rosario Fernández and J. M. Lassaletta from the Universidad de Sevilla and the CSIC de Sevilla, respectively. I also thank Prof. Tom Woo and Dr. Serge Gorelsky from the University of Ottawa for all the things they taught me during my short stay in Canada.

I would also like to thank the Universitat Autònoma de Barcelona for a UAB-PIF scholarship, and the Catalan Government for the 2009-BE1–00326 travel scholarship that made possible my short stay in the research group of Prof. Tom Woo at the University of Ottawa.

Finally, I would like to thank the most important people in my life, my family, because without them nothing could be possible.

Contents

Abbreviations

Ar	Any aromatic group
B3LYP	Becke's three-parameter, Lee-Yang-Parr exchange-correlation functional
^{t}Bu	*tert*-butyl
dba	dibenzylideneacetone
DCM	Dichloromethane
DMF	N,N-Dimethylformamide
DFT	Density Functional Theory
DMSO	Dimethyl sulfoxide
ee	Enantiomeric excess
EDG	Electron Donating Group
EWG	Electron Withdrawing Group
GGA	Generalized Gradient Approximation
HF	Hartree-Fock
IUPAC	International Union of Pure and Applied Chemistry
KS	Kohn-Sham
LDA	Local Density Approximation
LSDA	Local Spin Density Approximation
MA	Maleic Anhydride
Me	Methyl
MeCN	Acetonitrile
MM	Molecular Mechanics
MP2	Møller-Plesset perturbation theory to second order
NHC	N-Heterocyclic Carbene
NMR	Nuclear Magnetic Resonance
ONIOM	Our own N-layered integrated molecular orbital + molecular mechanics
PCM	Polarizable Continuum Model
PES	Potential Energy Surface
Ph	Phenyl
^{i}Pr	Isopropyl
QM	Quantum Mechanics
SDD	Stuttgart-Dresden effective core potential
S_N2	Bimolecular nucleophilic substitution

THF	Tetrahydrofuran
TOF	Turnover Frequency
TON	Turnover Number
ZPE	Zero Point Energy

Chapter 1
General Introduction

In all science, error precedes the truth, and it is better it should go first than last.

Hugh Walpole

As the title above suggests, this first chapter is devoted to make a general introduction to the main topic of this present thesis: the *Pd-catalyzed C–C cross-coupling reactions*. With this aim, in the first part of this chapter the concept of catalysis, its origin, as well as other related topics will be briefly introduced.

Next, in the second part of this chapter, the definition of C–C cross-coupling reactions and a summary of their generally accepted reaction pathway will be presented.

Finally, the elementary steps that take part in this type of reactions will be described separately. Furthermore, for each of these steps, the most relevant mechanistic studies on Pd-catalyzed C–C cross-coupling reactions reported in the last years, with particular attention to the theoretical ones, will be reviewed.

In principle, all the concepts provided in this first chapter should be more than enough to contextualize all the results obtained in the course of this thesis, and that are presented in Chaps. 4, 5 and 6.

1.1 Catalysis

1.1.1 Origin and Basic Concepts

The term "catalysis"[1] was introduced by Berzelius in 1836 when in an effort of accounting for a series of chemical transformations and decompositions, wrote: [1, 2]

[1] The word "catalysis" originates from the Greek words "κατά" and "λύσις" and has the sense of "dissolution".

M. G. Melchor, *A Theoretical Study of Pd-Catalyzed C–C Cross-Coupling Reactions*,
Springer Theses, DOI: 10.1007/978-3-319-01490-6_1,

"This new force, which was unknown until now, is common to organic and inorganic nature. I do not believe that this is a force entirely independent of the electrochemical affinities of matter; I believe, on the contrary, that it is only a new manifestation, but since we cannot see their connection and mutual dependence, it will be easier to designate it by a separate name. I will call this force catalytic force. Similarly, I will call the decomposition of bodies by this force catalysis, as one designates the decomposition of bodies by chemical affinity analysis".

Hence, Berzelius used the concept of catalysis to bring together a series of phenomena that could not be explained with the common conception at that time of reactions being caused by chemical affinity. The term, however, was originally purely descriptive because Berzelius intentionally refrained from attempting to explain the nature of catalysis. Specifically, he called "catalytic force" to a new force, but emphasized that he expected to be able to explain its operation within the framework of electrochemical theory.

At present, the International Union of Pure and Applied Chemistry (*IUPAC*) defines catalysis as *"the action of a substance (i.e. catalyst) that increases the rate of a reaction without modifying the overall standard Gibbs energy change in the reaction"*. In fact, this definition is quite similar to the one provided in 1895 by Ostwald[2]: *"a catalyst accelerates a chemical reaction without affecting the position of the equilibrium"*. Thus, a catalyst does not influence the thermodynamic equilibrium of reactants and products but affects the rates of the chemical reactions. In other words, the reaction Gibbs energy (ΔG) in a catalyzed reaction does not change whereas the activation energy (E_a) certainly does it (Fig. 1.1).

Nevertheless, catalysts can influence the reaction equilibria for more subtle reasons and, for example, result in higher concentrations of the product. It is well known that equilibria can be shifted by changing conditions, such as temperature, and that *Le Ch?telier's principle* predicts that an exothermic reaction will lead to higher equilibrium concentrations of product at lower temperatures than at higher temperatures. Thus, a catalyst that allows to conduct an exothermic reaction at a lower temperature than in its absence will lead to a higher concentration of product.

1.1.2 Transition State Stabilization

During a catalytic reaction, the catalyst decreases the activation energy, relative to that of the uncatalyzed reaction, by stabilizing preferentially the transition state(s). However, as the transition state is an unstable and fleeting species, the catalyst can not affect it alone. Instead, it coordinates to one or more of the reactants and remains coordinated throughout the transition state(s) of the catalytic process. Along this process, the dissociation of the product either regenerates the starting catalyst directly

[2] Wilhelm Ostwald was awarded with the Nobel Prize in Chemistry in 1909 *"in recognition of his work on catalysis and for his investigations into the fundamental principles governing chemical equilibria and rates of reaction"*.

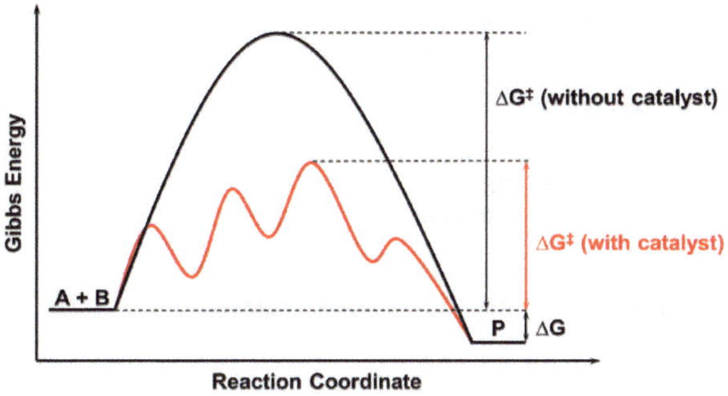

Fig. 1.1 General representation of an arbitrary reaction profile for a catalyzed (*red curve*) and uncatalyzed (*black curve*) reactions

or generates a species that will be converted to the starting catalyst. Consequently, catalysts can be used in sub-stoichiometric amounts relative to the reagents.

On the other hand, catalysts can lower the energy of the highest energy transition state by interacting with and stabilizing a structure that is similar to that of the uncatalyzed reaction, or can present a completely different reaction pathway (Fig. 1.1). The former is typical in enzyme-catalyzed processes while the latter occurs more often in organometallic chemistry. In this last case, the reaction normally takes place in more steps, but the activation energy of each of the individual steps is always lower than the activation energy of the uncatalyzed process.

1.1.3 The Catalytic Cycle

The combination of steps of a catalytic reaction is commonly known as *catalytic cycle*. As shown in Fig. 1.2, this series of reactions is generally depicted in the form of a cycle because the starting point of the catalytic process is also the ending point of the reaction (i.e. the catalyst is regenerated).

Often, the added catalyst is not in the active form. This is, the catalyst is a precursor and does not correspond to the species that enters into the catalytic cycle (Fig. 1.2). The species that runs in the catalytic cycle is generally named *catalytic species*, and can be generated either by the dissociation of a ligand or by the assistance of an *activator*. Once this active species is formed, it undergoes a series of transformations eventually giving rise to the formation of the final product and its concomitant regeneration. In some cases, however, during the catalytic cycle there can appear competitive reactions that give birth to inactive species, which produce the deactivation of the catalyst or the retard of the catalytic process. Examples of catalyst deactivation and retardation of the catalytic activity in Fig. 1.2 are the irreversible reaction of $[L_nM \cdot S]$ to form $[L_nM \cdot X]$, and the formation of the dimeric species $[L_nM{\cdot}S]_2$, respectively.

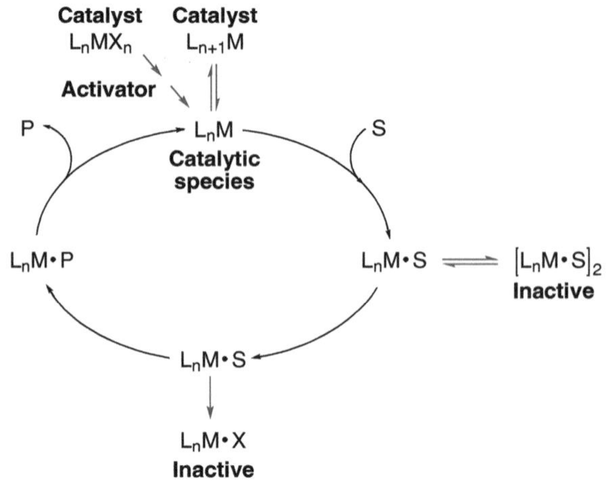

Fig. 1.2 Example of a catalytic cycle with both the reversible and irreversible formation of inactive species

1.1.4 Quantification of the Activity and Efficiency of Catalytic Processes

Even though in a catalytic reaction, in principle, the catalyst should be fully regenerated at the end of the process, it really rarely happens. As above mentioned, catalytic species can undergo competitive reactions and decrease their activity (i.e. catalyst deactivation). Thus, obtaining magnitudes that allow measuring their activity and efficiency is crucial for their comparison and the further design of more active and efficient catalysts.

A widely used manner of measuring the catalytic activity is to calculate the reaction rate constant, *k*. There exist different ways of obtaining this parameter. Probably, the simplest one is to use the empirical relation between the reaction rate constant and the temperature, derived by Arrhenius [3] in 1899 and that receives his own name[3]:

$$k = Ae^{\frac{-E_a}{RT}} \tag{1.1}$$

where *A* is the *pre-exponential factor* (sometimes called *frequency factor*), E_a is the *activation energy*, *R* is the *universal gas constant*, and *T* is the *absolute temperature* [4].

In the Arrhenius equation (1.1), the term *A* also changes with temperature. However, since its dependence compared to that of the exponential term is usually low, it can be neglected. On the other hand, the activation energy of a reaction is a very important term in catalysis because it dictates the barrier that must be overcome for a

[3] Although this expression applies only to gas phase reactions, it is often applied in general.

reaction to happen. The expression for this term can be easily obtained from Eq. (1.1) after simple mathematical treatment,

$$E_a = RT^2 \frac{d(ln\ k)}{dT} \tag{1.2}$$

and provides an empirical estimation of the energy barrier for a given reaction. A useful rule of thumb states that the reaction rate roughly doubles for every 10 °C increase in temperature. Moreover, in cases in which E_a is lower than ~5 kcal · mol^{-1}, we say that the reaction is diffusion-controlled, which means that the reaction rate is controlled by how fast the molecules move toward one another. Alternatively, when E_a is higher than ~5 kcal · mol^{-1}, we say that the reaction is chemically-controlled [4].

Another way of calculating the reaction rate constant is by means of the transition state theory (*TST*). This theory was developed simultaneously in 1935 by Eyring [5] and by Evans and Polanyi [6]. There exist different formulations of the *TST*, but only the thermodynamical formulation will be described herein.

Let us consider a reaction between two reactants, *A* and *B*, that results in an activated species $AB^{\ddagger}$ (or transition state) such that it spontaneously evolves to the final product *P*, and further, is in equilibrium with the starting reactants:

$$A + B \underset{k_{-1}^{\ddagger}}{\overset{k_1^{\ddagger}}{\rightleftharpoons}} AB^{\ddagger} \overset{k^{\ddagger}}{\longrightarrow} P \tag{1.3}$$

The equilibrium constant for the reaction between reagents *A* and *B*, and the activated species $AB^{\ddagger}$ is given by:

$$K_c^{\ddagger} = \frac{k_1^{\ddagger}}{k_{-1}^{\ddagger}} = \frac{[AB^{\ddagger}]}{[A][B]} \tag{1.4}$$

The concentration of the activated species $[AB^{\ddagger}]$ is considered to be negligible towards the concentration of reagents *A* and *B*. Thus, the concentration of the reagents can be assumed constant. On the other hand, the reaction rate can be expressed as a function of the formation of product *P*,

$$\frac{d[P]}{dt} = k^{\ddagger}[AB^{\ddagger}] \tag{1.5}$$

where $k^{\ddagger}$ is the rate constant for the reaction going from $[AB^{\ddagger}]$ to *P*. The value of this rate constant can be obtained by statistical mechanics and equals to $k_B T/h$, being k_B the *Boltzmann constant* ($1.38 \cdot 10^{-23}$ J · K^{-1}), and *h* the *Planck constant* ($6.626 \cdot 10^{-34}$ J · s).

Now, if we combine the Eqs. (1.5) and (1.4) we arrive at:

$$\frac{d[P]}{dt} = \underbrace{\frac{k_B T}{h} K_c^{\ddagger}}_{k} [A][B] \tag{1.6}$$

Finally, the *van't Hoff equation* states the relation between the equilibrium constant and the variation in the *activation Gibbs energy* ($\Delta G_0^{\ddagger}$) by means of the following expression,

$$\Delta G_0^{\ddagger} = -RT \, ln \, K_c^{\ddagger} \tag{1.7}$$

where $\Delta G_0^{\ddagger}$ is defined as the difference between the standard Gibbs energy of the transition state $AB^{\ddagger}$ and the reagents A and B. Introducing the value of $K_c^{\ddagger}$ in Eq. (1.7) into the Eq. (1.6) we obtain that the reaction rate constant of the overall reaction is:

$$k = \frac{k_B T}{h} e^{-\frac{\Delta G_0^{\ddagger}}{RT}} \tag{1.8}$$

In this last equation $\Delta G_0^{\ddagger}$ can be alternatively expressed in terms of enthalpy and entropy of activation ($\Delta H_0^{\ddagger}$ and $\Delta S_0^{\ddagger}$, respectively), which leads to the widely known *Eyring equation*:

$$k = \frac{k_B T}{h} e^{-\frac{\Delta H_0^{\ddagger}}{RT}} e^{\frac{\Delta S_0^{\ddagger}}{R}} \tag{1.9}$$

So far, we have seen that the reaction rate constant is used to quantify the catalytic activity and how it can be calculated. But, how is the efficiency of a catalytic process measured? In general, two quantities are used for this purpose. The first one is the *turnover number*, *TON*. As its name suggests, it can be seen as the number of turns that a catalyst does within the catalytic cycle before it is deactivated. It can be calculated as the number of moles of product per mole of catalyst, as shown in Eq. (1.10). Hence, for an ideal catalyst, the *TON* should be infinite. So that you can get an idea of the range of *TON* values, commodity-scale reactions such as rhodium-catalyzed hydroformylation occur with hundreds of thousands of turnover numbers, whereas other reactions that are less developed might occur with 10–20 turnovers [7].

$$TON = \frac{moles \ of \ product}{moles \ of \ catalyst} \tag{1.10}$$

The other quantity used to measure the efficiency of a catalytic process is the *turnover frequency*, *TOF*, which is defined as the number of moles of product per mole of catalyst per unit time. Therefore, the *TOF* is simply the *TON* per unit time:

$$TOF = \frac{TON}{time} \tag{1.11}$$

The above presented definitions for the *TON* and *TOF*, however, vary slightly depending on the type of catalyst. In heterogeneous catalysis, for example, the *TON* and *TOF* are often defined per active site or per gram catalyst. In contrast, in biocatalysis, they are defined by the rate measured when all the enzyme molecules are complexed with a reactant, divided by the total enzyme concentration. Thus, when dealing with *TON* and *TOF* data it is important to provide the units in order to avoid misunderstandings.

Up to a few years ago, there was no simple way to calculate the efficiency of a catalytic cycle (i.e. *TOF*) from a theoretically obtained energy profile. It was Kozuch and Shaik who recently introduced the so-called *energetic span model*, which enables evaluating *TOFs* in a straightforward manner [8–11]. The formulation of this model is based on *Eyring's TST* and corresponds to a steady-state regime. One of the most interesting conclusions derived by the authors from this model is that in catalytic cycles there are no rate determining steps, but *rate determining states*.

With the concept of catalysis and some of their related aspects already introduced, the following section will be devoted to describe the type of catalytic reactions that have been studied in this present thesis: the *C–C cross-coupling reactions*.

1.2 C–C Cross-Coupling Reactions

1.2.1 What Are C–C Cross-Coupling Reactions?

The C–C cross-coupling reactions are undoubtedly one of the most important and useful reactions in organic synthesis and organometallic chemistry, since a huge variety of complex compounds can be synthesized from readily accessible reactants [12–16]. These reactions consist in the carbon-carbon bond formation between an organic electrophile (R^1–X) and an organometallic nucleophile (R^2–m) in the presence of a metal catalyst [M] (Scheme 1.1). Generally, the most widely used metal catalysts are transition metal complexes from groups 8–10, especially Ni and Pd [17–19]. The reason for which these metals are the most common ones originates from their Ni(II)/Ni(0) and Pd(II)/Pd(0) ease of redox exchange, which is an indispensable condition for the catalytic cycle to be completed. Between these two catalysts, however, Pd catalysts have been demonstrated to have more advantages for cross-coupling reactions than those of Ni [7]. For example, they tend to be less sensitive to oxygen and are believed to be less toxic. Moreover, unlike Ni catalysts, Pd complexes tend to react without the intervention of radical intermediates, which can eventually lead to side products such as those from homocoupling.[4] Consequently, Pd-catalyzed cross-coupling reactions have been the most developed ones.

As regards to the organometallic nucleophile (R^2–m), depending on the metal or semi-metal involved, cross-coupling reactions receive different names, which usu-

[4] The *homocoupling* is the coupling of two identical molecules.

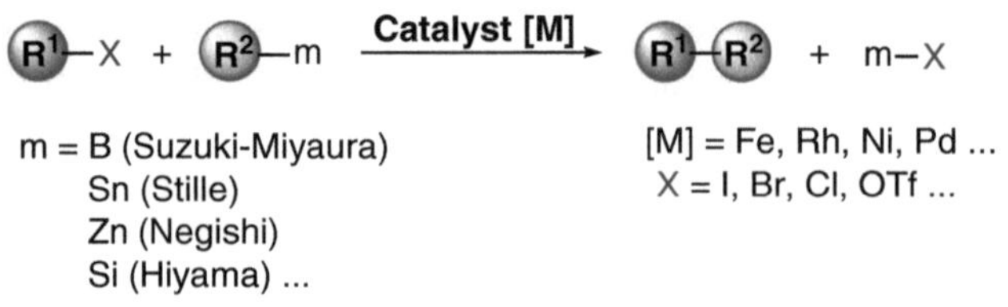

Scheme 1.1 General scheme for C–C cross-coupling reactions

ally correspond to their discoverers' name (Scheme 1.1). For instance, Stille [20–22] reaction is tin-mediated; Suzuki-Miyaura [23, 24], boron-mediated; Negishi [25], zinc-mediated, etc. The relevance of these reactions in the last decades has been such that three of their developers were awarded with the Nobel Prize in Chemistry in 2010.[5] Additionally, their mild reaction conditions along with their wide tolerance of functional groups have substantially contributed to their application in many areas, such as the synthesis of natural products [26–28], fine chemicals [29–31], and polymers [32–34].

Among the numerous advantages of these reactions, probably the most remarkably one is the manifold variety of organometallic reagents and organic electrophiles that can be employed to assemble C–C bonds. However, not all the reactions involving different organic groups (e.g. R = alkyl, allyl, aryl, alkenyl, alkynyl) and different heteroatoms (e.g. X = Cl, Br, I) have been studied to the same extent. For example, cross-coupling reactions involving aryl or vinyl electrophiles have been by far the most extensively studied mainly because they are readily accessible reagents and, further, their reactions have wide applicability and high selectivities. Conversely, alkyl electrophiles, and specially those possessing β-hydrogens (i.e. unactivated alkyl electrophiles), have been much less studied. In this case, it is mainly due to their lack of reactivity towards oxidative addition and their ease of participating in competitive reactions such as β-elimination and hydrohalogenation. In fact, it was not until the appearance of some pioneering works, such as the ones from Kochi and Tamura [35–37], Suzuki et al. [38], and other recent works, such as the ones from Knochel [39–42] or Beller [43], that it was demonstrated that cross-coupling reactions employing alkyl electrophiles are feasible. Thereafter, the interest in these reactions has become increasingly important to the point that the development of efficient Csp^3–Csp^3 coupling is at present one of the active research topics within cross-coupling reactions.

On the other hand, cross-coupling reactions involving organic halides initially had the important limitation that only aryl bromides and iodides could be employed. However, as aryl chlorides are more profusely available and, in general, less expensive

[5] Richard F. Heck, Ei-ichi Negishi and Akira Suzuki shared the Nobel Prize in Chemistry in 2010 *"for palladium-catalyzed cross couplings in organic synthesis"*. These scientists are the developers of the widely known Heck, Negishi, and Suzuki reactions, respectively.

than their bromide and iodide analogs, many efforts to overcome that issue have been made since then [44]. In particular, important advances in the past few years have been achieved mainly owing to the development of new transition metal complexes containing electron rich and bulky ligands (mostly phosphines [45–47], but also carbenes [48, 49]).

Recently, direct C–H arylation [50–54] has emerged as an elegant and effective alternative to C–C cross-coupling reactions since these reactions do not require the presence of the organometallic nucleophile and provide only HX associated to a base as by-product. Therefore, these reactions are very interesting both in terms of atom-economy and the relative toxicity of the wastes.

1.2.2 The Catalytic Cycle

C–C cross-coupling reactions follow a general accepted reaction pathway that consists of three steps (Fig. 1.3): *(i) oxidative addition* or insertion of the low valent transition metal into the electrophilic carbon-heteroatom bond; *(ii) transmetalation* or displacement of the heteroatom leaving group by the nucleophile; and finally, *(iii) reductive elimination* to form the new C–C bond with the concomitant regeneration of the catalyst. The former and the latter steps are common to all cross-coupling reactions and have been studied in depth both by experimental [55–58] and computational methods [59–62]. Hence, the mechanisms for these two steps are quite well understood. In contrast, the various cross-coupling reactions differ in the nucleophile used for transmetalation and, consequently, in the transmetalation step. Furthermore, experimental evidence for this process is particularly difficult to obtain due to the complexity that involves the isolation and characterization of key intermediates. Thus, it is not surprising that the transmetalation step is the less studied and considered as the most complex one within the catalytic cycle (Fig. 1.3). Even so, in the last years several mechanistic studies on this step have appeared [63–69]. In particular, in our group we have carried out theoretical studies concerning this step in several cross-coupling reactions, some of them in collaboration with experimental groups [70–77]. In fact, some of these theoretical studies are part of the results of this thesis and will be presented in Chaps. 4, 5 and 6.

In the following sections, a brief description of the three elementary steps of the catalytic cycle (Fig. 1.3) will be presented separately. More specifically, some of the most relevant studies reported in the last years on these steps will be reviewed.

1.2.3 Oxidative Addition

Oxidative addition reactions are processes in which the bond in an organic or main group reagent A–B breaks, and two new bonds with the metal are formed (Scheme 1.2). In this overall process, since A and B are anionic ligands, the metal is formally

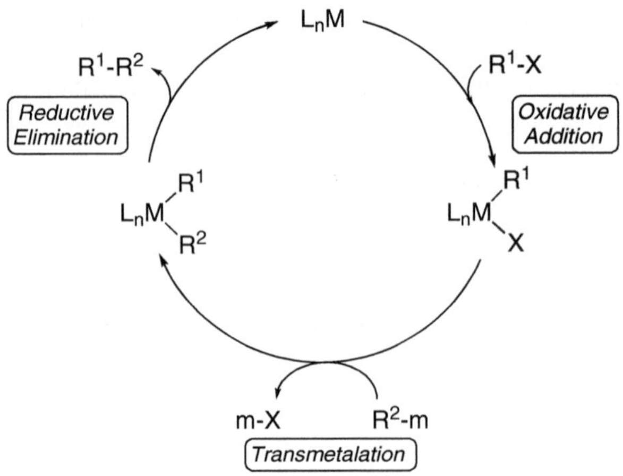

Fig. 1.3 General catalytic cycle for C–C cross-coupling reactions

oxidized. More specifically, oxidative additions to mononuclear metal complexes (Eq. 1, Scheme 1.2) results in an increase in the oxidation state of the metal from n to n+2, whereas oxidative additions to dinuclear metal complexes (Eq. 2, Scheme 1.2) increase the oxidation state of each metal from n to n+1.

The opposite reaction to the oxidative addition is the reductive elimination, where the A–B molecule is expelled from the [M(A)(B)] complex. In principle, these reactions can be reversible but, in practice, they tend to evolve in either one or other direction. In fact, the position of equilibrium in any particular case depends on the overall thermodynamics, which in turn depends on the relative stability of the metal in each oxidation state and the strength of the A–B bond with respect to the M–A and M–B bonds. On the basis of these dependences and other chemical concepts (e.g. coordination number of the metal), we can consider the following series of trends as a guide for predicting the reactivity of metal complexes towards oxidative addition[6]:

- *Oxidative additions to more electron-rich metal centers tend to be more favorable than oxidative additions to more electron-poor metal centers.*
- *Oxidative additions to less-hindered metal centers tend to be more favorable than oxidative additions to more-hindered metal centers.*
- *Rates and equilibrium constants for ligand dissociation or association previous to oxidative additions affect the rates of the overall addition processes.*

Probably, one of the best features of oxidative addition reactions is the unusual wide range of reagents A–B that can be involved. These can be divided into three groups: *(i)* species that are non-polar or have low polarity (e.g. H_2, silanes); *(ii)* reagents that are highly polar (e.g. alkyl halides, strong acids); and *(iii)* reagents that are intermediate in polarity (e.g. amines, alcohols). A direct consequence derived

[6] These trends are only given by way of guidance; there may be exceptions.

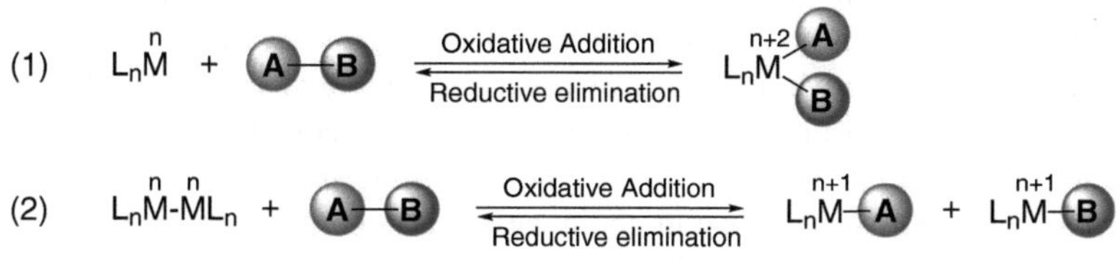

Scheme 1.2 General scheme for the oxidative addition reaction of an A–B molecule to a mononuclear (Eq. 1) and dinuclear (Eq. 2) metal complexes

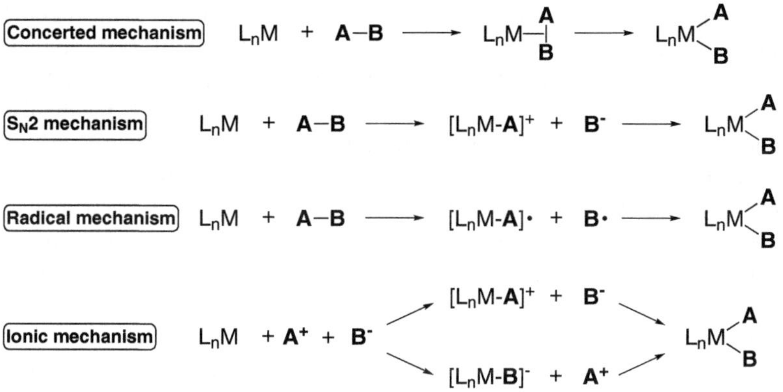

Fig. 1.4 Reaction mechanisms for the oxidative addition of an A–B molecule to a metal complex

from this wide range of A–B molecules that can undergo oxidative addition is the existence of multiple reactions mechanisms for this process. In particular, in the literature there are described four reaction mechanisms for oxidative addition (Fig. 1.4): *concerted mechanism*, S_n2 *mechanism*, *radical mechanism*, and *ionic mechanism*.

In oxidative additions via a *concerted mechanism* (Fig. 1.4), the A–B molecule binds firstly to the metal center and then, the cleavage of the A–B bond and the formation of the new M–A and M–B bonds take place simultaneously through a three-centered transition state. This mechanism is normally found in oxidative additions of non-polar reagents [78–81] and aryl halides [82–85]. Experimental evidences for this mechanism are the retention of configuration at a stereogenic center in the case of chiral A–B reagents, and the relative *cis* disposition of the ligands A and B after the oxidative addition [86]. The latter, however, may be not observed in the cases in which the cis-to-trans isomerization reaction of the oxidative product is very fast [87].

On the other hand, the S_n2 *mechanism* is an associative bimolecular process that consists in two steps (Fig. 1.4): first, the ligand A is attacked by a metal electron pair and the anionic ligand B^- is expelled giving rise to the cationic species $[L_nM–A]^+$; subsequently, the two charged species collapse to yield the oxidative addition product. Unlike the concerted mechanism, the relative position of the ligands A and B in the final product via this mechanism can be either *cis* or *trans*, depending on where the anionic ligand B ends up coordinated after the second step.[7] This S_n2 *mechanism* is often found in oxidative additions of polar reagents and in polar solvents, and results in the inversion of configuration of a stereogenic center [86, 88–90].

The third mechanism in Fig. 1.4 is a type of *radical mechanism* known as *non-chain radical mechanism*.[8] This non-chain variant is generally believed to operate in oxidative additions of certain alkyl halides [92] and consists in the one-electron oxidation of the metal by the A–B molecule giving rise to the radical species $[L_nM–A]\cdot$ and $B\cdot$, which rapidly recombine to yield the oxidative addition product $[L_nM(A)(B)]$. Similarly to the S_n2 *mechanism*, the rate of this radical mechanism increases as more basic is the metal, and the more easily the ligand A is transferred to the metal. Experimental evidences for this type of mechanism can be the significant changes in the reaction rate produced by the introduction of slight modifications of the substrate, the metal complex, or the solvent. Another alternative to confirm this mechanism is to use radical scavengers, such as $RNO\cdot$. This alternative, however, has been somewhat criticized on the basis that the presence of such species may initiate a radical pathway for a reaction that otherwise would have followed a non-radical mechanism.

Finally, the last mechanism proposed for oxidative addition is the *ionic mechanism* (Fig. 1.4). This mechanism is adopted for A–B molecules that are completely dissociated in solution, and consists in the consecutive coordination of the two charged ligands to the metal complex. Thus, this mechanism can evolve through two possible variants. The first one involves the initial coordination of the cation A^+ to the metal complex and the subsequent coordination of the anionic ligand B^-. On the contrary, in the second alternative the coordination of the anionic ligand B^- to the metal complex occurs first, followed by the coordination of the cation A^+. In general, the first variant is the most common one and is favored by basic ligands and a low oxidation state of the metal [91].

Overall, we have seen that the reaction mechanism for the oxidative addition is highly sensitive to the nature of the reagent A–B. Thus, depending on the reagent one or another mechanism might be favored. However, this is not the unique factor that influences the oxidative addition. Other factors that can affect this process are, for instance, the nature of the solvent, the metal-bound ligands or the added additives. An illustrative example of this is the oxidative addition of allylic chloride to Pd(0) and Pt(0) complexes [93, 94]. In these reactions, the use of polar solvents, such as *MeCN* or *DMSO*, was found to favor the S_n2 *mechanism* leading to complete or nearly complete inversion of configuration. Conversely, complete or predominant retention of configuration was observed in less coordinating solvents, such as benzene or dichloromethane.

---

[7] This may be governed by the *trans-effect*.
Trans-effect: is the effect that produces a ligand that facilitates the exchange or substitution of a second ligand that is in *trans* position respect to the former. We refer to ligands with high trans-effect when these facilitate that substitution.

[8] There is also another variant of radical mechanism called *chain mechanism*. This, however, will not be discussed here. More details on this type of mechanism can be found in books on general organometallic chemistry [7, 91].

Among the manifold variety of A–B molecules that can undergo oxidative addition, the ones possessing C–X bonds are typically involved in the oxidative addition step of cross-coupling reactions. Hence, since this thesis is focused on the theoretical study of these reactions catalyzed by Pd complexes, I found appropriate to include at the end of this section a brief overview of this type of oxidative additions. The aim of this overview is just to provide a general picture of which have been the main research interests related to this process, paying especial attention on the reported theoretical works.

Oxidative Addition of C–X Bonds to Pd(0) Complexes

Oxidative additions of substrates possessing C–X bonds, and specially those involving Ar–X species, have been extensively studied because they are the first step in cross-coupling reactions. The order in reactivity of the different organic halides is the following: I > Br > Cl [95]. Thus, the focus of computational attention on this step can be attributed to the fact that it has been postulated to be rate-limiting in a few cross-coupling reactions, specially when X = Cl [44].

One of the earliest theoretical studies on C–X oxidative addition to a Pd complex was reported by Bickelhaupt et al. [81]. These authors investigated the addition of chloromethane to a ligand-free palladium atom by means of density functional theory (*DFT*) calculations showing a preference for the concerted pathway toward the S_n2 *substitution*. Few years later, the oxidative addition of Ph–X (X = Cl, Br, I) to Pd(0) complexes containing bidentate phosphanes was investigated by Senn and Ziegler [59]. Interestingly, the authors could not locate in solution the concerted transition state reported for the gas phase. Instead, they found that the dissociation of the halide and its subsequent recombination with the cationic phenyl complex (i.e. the S_n2 *mechanism*) was energetically a facile process in solution even without accounting for favorable entropic contributions that such a dissociative mechanism may have.

As previously stated, the oxidative addition reaction is highly sensitive to the nature of the reagent, the added additives, the metal-bound ligands, and the solvent. Hence, all these factors have also been of particular interest in this type of oxidative additions and have been studied in some depth. Regarding the influence of added additives, the role of anionic additives in improving the oxidative addition step has been analyzed by several authors. Jutand et al. [55, 96, 97] provided a significant amount of experimental evidences suggesting an anionic form for the catalyst, $[Pd(Cl)(L)_n]^-$. Calculations by Bickelhaupt and co-workers on the activation of H–H, C–H, C–C and C–Cl bonds showed that the anionic species $[PdCl]^-$ favors the S_n2 *pathway* even in gas phase, and revealed that the lowering of activation barriers through anion assistance is caused by the more stabilizing interaction between the reactants in the transition state [98]. On the other hand, Thiel et al. [60] explored the reaction of $[Pd(OAc)(PMe_3)_2]^-$ with Ph–I. In this case, however, the mechanism is quite subtle; the acetate ligand moves away from palladium when iodobenzene coordinates, and returns afterwards to the metal to displace another ligand. Two different

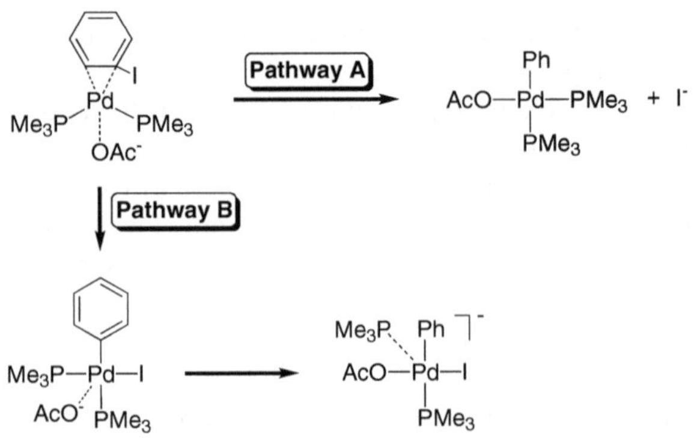

Fig. 1.5 Reaction pathways proposed by Thiel et al. for the oxidative addition of Ph–I to $[Pd(OAc)(PMe_3)_2]^-$ [60]

paths of similar energies were characterized for this process (Fig. 1.5): the most favored one (pathway A, Fig. 1.5) ends up with departure of iodide and resembles the S_n2 *pathway*; the least favored one (pathway B, Fig. 1.5), involves the displacement of a phosphine ligand and the transition state associated to C–I cleavage is of a concerted type.

As regards to the influence of the nature of the substrate, so far we have seen that the prototypical substrates for oxidative addition reactions are aryl halides, but fragments different from aryl are also relevant. Aimed at providing a better understanding of what is responsible for the higher reactivity of aryl halides compared to alkyl halides, Ariafard and Lin recently reported a *DFT* study on the oxidative addition of different reagents R–Br (R = methyl, benzyl, phenyl, vinyl) to $[Pd(PH_3)_2]$ [99]. In this study, however, the authors only considered the oxidative addition reaction via a concerted pathway. As expected, the calculated barriers with the systems involving sp^2 carbons resulted to be lower than those involving sp^3 carbons. In particular, the Gibbs energy values in gas phase increased in the order: vinyl > phenyl > benzyl > methyl. The authors attributed this order in the reactivity to kinetic factors rather than thermodynamic factors. More specifically, they demonstrated that the lowering of the reaction barriers is consequence of the existence of low-lying C–Br π^*-orbitals in the transition states structures of the unsaturated systems.

Last but not least, the effect of the donating character of the ligands and the number of them bound to the metal center in the active form of the catalyst, have been two of the factors that have drawn more attention. Usually, in palladium systems, the catalyst is introduced as a precursor, often as $[Pd_2(dba)_3]$ or $[Pd(PPh_3)_4]$; however, this is clearly not the active form. The bis-ligated form has been postulated in most experimental proposals, and accordingly used in calculations, because of the well established stability of d^{10} $[ML_2]$ complexes, but the mono-ligated form [ML] can be also envisaged (Fig. 1.6).

In fact, the enhanced reactivity of transition metal complexes containing electron-rich and bulky phosphine ligands has been attributed to the formation of mono-

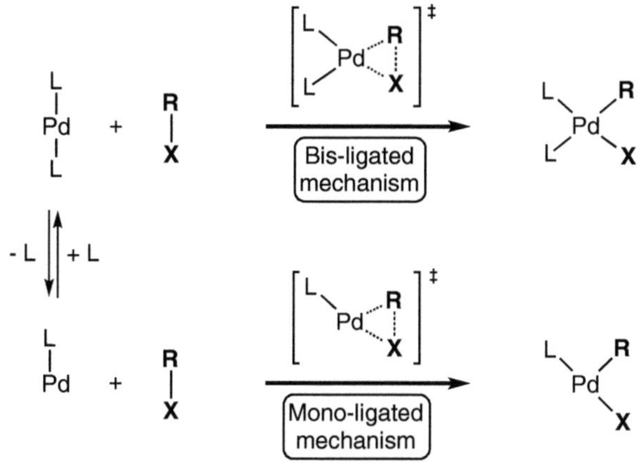

Fig. 1.6 Bis-ligated and mono-ligated reaction pathways proposed for the oxidative addition of R–X to [PdL_2] complexes

ligated [ML] species, which can undergo oxidative addition more rapidly than the corresponding bis-ligated ones [ML_2] [44, 100, 101]. On this issue, several authors have recently gone even further and have revealed that the rate constants for the oxidative addition depend more on the identity of the halide in the Ar–X reagent rather than on the steric bulk of the phosphine ligands [102–104]. In particular, these authors independently came to the same conclusion that the use of bulky ligands for the oxidative addition of Ar–Cl and Ar–Br promotes the reaction because such ligands favor the mono-ligated pathway, while Ar–I species are more susceptible to oxidative addition and therefore, they favor the bis-phosphine pathway.

More recently, the electronic and steric effects of different phosphine ligands (i.e. PF_3, PH_3, PMe_3 and PPh_3) was theoretically investigated by Maseras et al. for the oxidative addition of CH_3Br to mono- and bis-ligated palladium complexes through both concerted and S_n2 *mechanisms* [105]. According to the authors, steric effects seem to be critical in the concerted mechanism for the bis-ligated systems, where the energy barrier resulted to increase with the bulk of the phosphine. On the contrary, the phosphine effect in the other cases was found to be mostly electronic, with the barrier being lower with more electron-donating phosphines.

Besides all the above mentioned works, there are many other theoretical studies related to the effect of phosphine ligands on cross-coupling reactions, but for the lack of space and time only some of them have been cited herein. Nevertheless, if you have further interest, you can look up our book chapter [106] in which most of the relevant theoretical studies on phosphine effects in the different steps of some Pd-catalyzed cross-coupling reactions are reviewed.

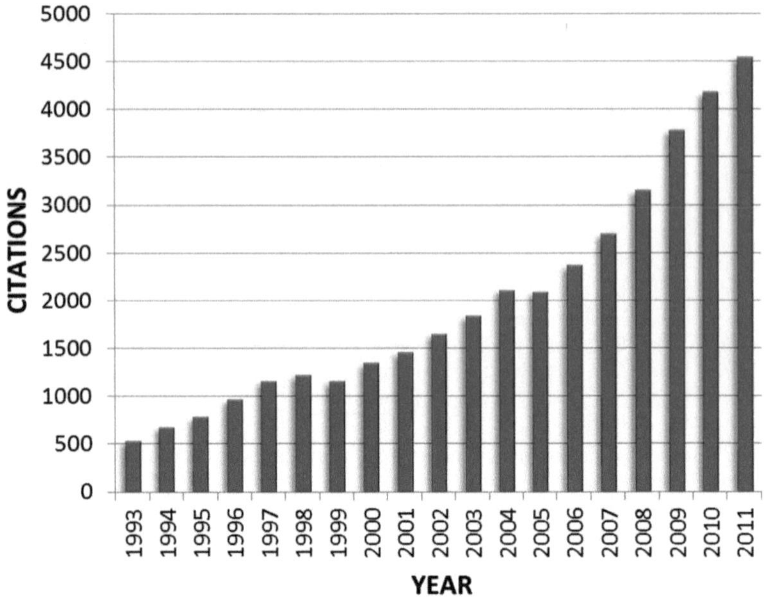

Fig. 1.7 Plot of the number of citations per year between 1993 and 2011 for the topic "transmetalation". Data from the *Web of Knowledge* database (December 2011)

1.2.4 Transmetalation

Transmetalation reactions consist in the transfer of an organic group R (e.g. alkyl, aryl, vinyl) from one metal (M^1) to another one (M^2) [107–109]. Thus, these reactions entail the cleavage of a M^1–C bond and the concomitant formation of a new M^2–C bond. Regarding the metal centers, these can be both transition metals and main group elements. As a result, depending on the type of the metal atoms involved in transmetalation we can distinguish between three different processes: *(i)* organic ligand transfer between two main group elements; *(ii)* that between one main group element and one transition metal; and *(iii)* transmetalation involving two transition metals.

Compared to other fundamental organometallic reactions involving transition metal complexes (e.g. oxidative addition), transmetalation reactions have been much less studied to the point that have even been considered simply as ligand substitutions. Recently, however, as a consequence of its participation in relevant catalytic processes (e.g. C–C cross-coupling reactions), they have started being more studied and, at present, are recognized as a differentiated process. A prove of its increasingly importance is the number of citations per year in scientific journals for *"transmetalation"* as topic (Fig. 1.7).

A common classification for transmetalation reactions is the one based on the reaction mechanisms, depicted in Fig. 1.8. In the first category of this classification we

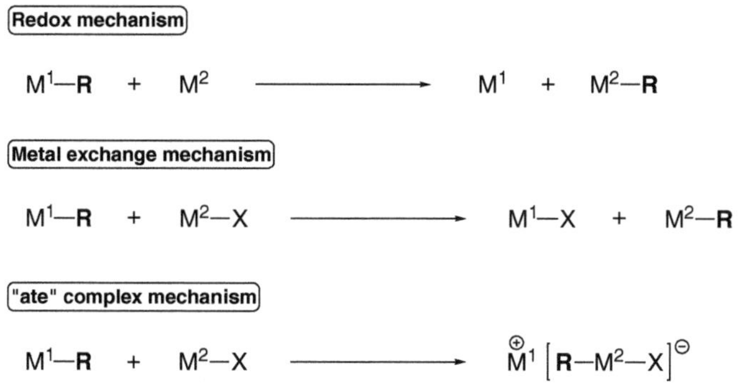

Fig. 1.8 Mechanisms proposed for transmetalation reaction

can find those transmetalations that follow a *redox-type mechanism*, which consists in the intermolecular transfer of the organic ligand accompanied by the oxidation and reduction of the metal centers involved. The thermodynamics of this type of transmetalation reactions will be governed by the relative stability of the metal-carbon bonds of the started and produced organometallic compounds.

In the second category of that classification we have those transmetalation reactions that evolve through a *metal exchange mechanism* (Fig. 1.8). As its name suggests, this mechanism entails the metal-ligand exchange between an organometallic compound M^1–R and a metal complex with a halogen or pseudo-halogen ligand M^2–X (e.g. X = Cl, Br, CN). This type of transmetalation is the most common among the three types of reactions and is characteristic of cross-coupling reactions. On the other hand, the exchange reaction of alkyl or aryl ligands between two organometallic compounds (i.e. X = R') is also included within this category.

Lastly, within the third type of transmetalation reactions we find those that follow a mechanism dubbed as *"ate" complex mechanism* (Fig. 1.8). In this mechanism, the transfer of the organic ligand from M^1 to M^2 gives birth to an ion-pair formed between the metal cation and the anionic organometallic complex. This mechanism differs from the metal exchange mechanism in that it only involves the transfer of the organic group from M^1 to M^2, and not the simultaneous intermolecular transfer of both the organic and the halogen (or pseudo-halogen) ligands between M^1 and M^2.

Given that the carbon-bonded ligand does not easily dissociate, all the above mentioned reaction mechanisms involve transition states (or intermediates) where the organic group acts as a bridging ligand connecting the two metal centers (Fig. 1.9). Additionally, in some exchange mechanisms the halide or pseudo-halide can also act as stabilizing bridging ligand in the transition state. In such cases, the transition states are usually called *"cyclic"*, whereas the rest are referred to as *"open"*. On the other hand, the formation of these bimetallic species and its dissociation into two metal complexes via a concerted pathway suggest that transmetalation reactions may be reversible.[9] However, the reversibility or not of these reactions might be dictated

[9] The reverse of the transmetalation reaction is commonly known as "retrotransmetalation" [74].

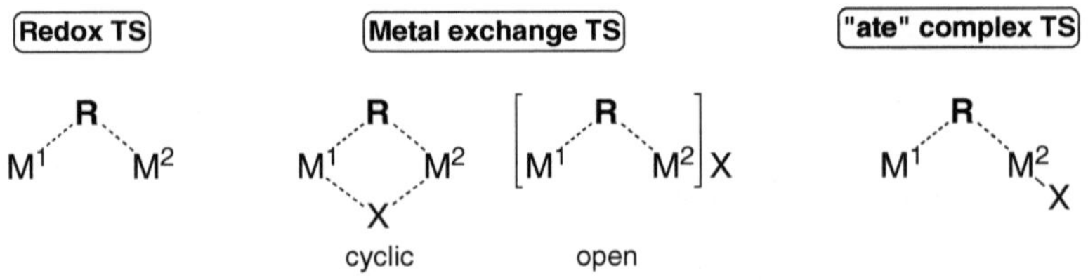

Fig. 1.9 Schematic representation of the transition states proposed for the different types of transmetalation reactions

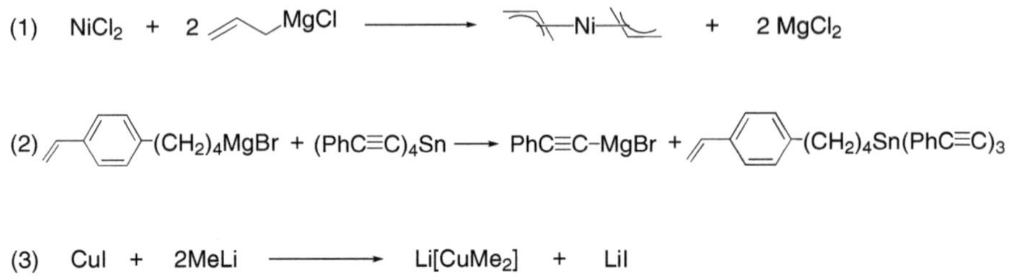

Fig. 1.10 Examples of transmetalation reactions used for the synthesis of organometallic compounds that involve Grignard and organolithiate reagents [91]

by the influence of several factors, such as the reaction rate of the reaction following transmetalation, the nature of the ligands bound to the metal complex, or the nature of the organic group. Thus, the reversibility of these reactions needs to be evaluated for each particular case.

As far as the applications of transmetalation reactions is concerned, these reactions are very important; on the one hand, because they are employed for the synthesis of new organometallic complexes, and on the other hand, because they are part of catalytic cycles for preparing organic compounds. In particular, within the transmetalation reactions used for the synthesis of organometallic compounds there are included all the reactions in which halides (or pseudo-halides) of transition metals (or main group elements) react with organometallic complexes of main groups elements to yield new organometallic complexes. The most common reagents that act as a alkylating agents of other metals are the Grignard reagents and the organolithiates [91]. Some examples for these reactions are shown in Fig. 1.10. Interestingly, unlike most of the Grignard reagents that react through metal-exchange reactions (Eqs. 1 and 2, Fig. 1.10), organolithiates are stronger alkylating agents and thus, they often undergo transmetalation via an "ate" complex mechanism (Eq. 3, Fig. 1.10).

Among the different roles of transmetalation within catalytic cycles, its role within cross-coupling reactions is, perhaps, the most relevant one. In fact, the different cross-coupling reactions differ in the organometallic nucleophile used and, accordingly, in the transmetalation step (Scheme 1.3). This step takes place from the oxidative addition product *cis*- or *trans*-$[M(R^1)(X)(L)_n]$ and results in the formation of the complex $[M(R^1)(R^2)(L)_n]$, where the two organic groups can also be in a *cis* or

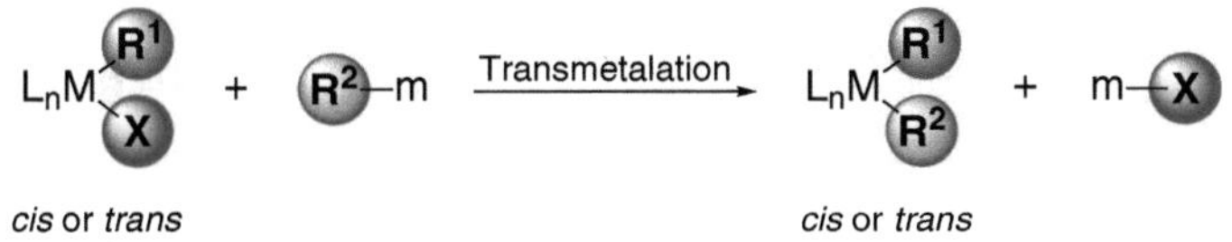

Scheme 1.3 General scheme for the transmetalation process in C–C cross-coupling reactions

trans disposition. Importantly, the subsequent reductive elimination step must occur from the *cis* isomer. Hence, in the cases in which transmetalation leads to the *trans* product, the trans-to-cis isomerization has to take place in order to successfully complete the catalytic cycle. On the other hand, the product of the oxidative addition is commonly assumed to be *trans* because this isomer is more stable than the *cis* one and, further, the cis-to-trans isomerization is usually fast. However, when this is not the case and that isomerization is not fast, both *cis* and *trans* isomers can compete for transmetalation and affect the final outcome of the reaction. Therefore, in such cases, the transmetalation reactions starting from both the *trans* and *cis* isomers have to be considered. An example of this situation was exposed meticulously and with an expert hand in a recent experimental work on the Stille coupling of alkynyl stannane and aryl iodide reported by Espinet et al. [110].

Similarly to oxidative addition, the transmetalation step can be also influenced by slightly modifications on the reaction conditions (e.g. nature of the R group, metal-bound ligands, solvent). In this case, however, this influence is not so straightforward as in the oxidative addition. This is mainly due to the fact that, since cross-coupling reactions differ in the transmetalation step, the reaction mechanisms proposed for these processes are more varied and complex. Thus, only the reaction mechanisms proposed for the transmetalation step of the cross-coupling reactions that have been studied in this thesis will be presented. Specifically, they will be presented in the introduction of the chapter of the corresponding cross-coupling reaction.

1.2.5 Reductive Elimination

Reductive elimination is the reverse of oxidative addition. Thus, in the overall process, two metal-ligand bonds are broken and one new ligand-ligand bond is formed (Scheme 1.4). This coupling of two covalent ligands can further take place both at a single metal center (Eq. 1, Scheme 1.4) and between two ligands from two different metal centers (Eq. 2, Scheme 1.4). In any case, as expected for the reverse process of the oxidative addition, the oxidation state of the metal is reduced. More specifically, it is reduced by two units in the case of mononuclear metal complexes, and by one in each of the two metal centers in the case of dinuclear metal complexes.

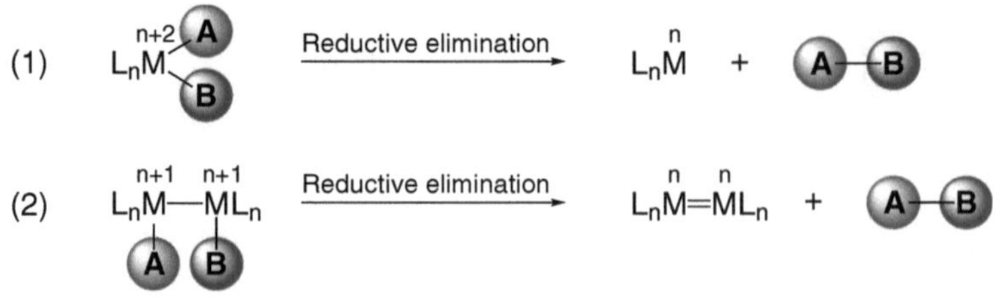

Scheme 1.4 General scheme for the reductive elimination of an A–B molecule from a mononuclear (Eq. 1) and dinuclear (Eq. 2) metal complexes

Similarly to oxidative addition, there are many factors that control the rates of reductive elimination reactions. In many cases, some of these factors have the opposite effect on the rate of reductive elimination than in the oxidative addition because these effects originate from thermodynamic factors. Thus, those factors that thermodynamically favor oxidative addition must thermodynamically disfavor the opposite reductive elimination reaction. On this basis, we can consider the following series of trends as a guide for predicting the reactivity of metal complexes towards reductive elimination[10]:

- *Reductive eliminations in first row metal complexes tend to be more favorable than in second row metal complexes, which tend to be more favorable than in third row metal complexes.*
- *Reductive eliminations in more electron-poor complexes tend to be more favorable than in more electron-rich complexes.*
- *Reductive eliminations in more-hindered complexes tend to be more favorable than in less-hindered complexes.*
- *Reductive eliminations of H ligands are faster than those of R ligands.*

As regards to the reaction mechanisms involved in this type of reactions, the principle of microscopic reversibility[11] suggests that reductive elimination reactions should show the same variety of mechanisms than oxidative additions (Fig. 1.4). Furthermore, it implies that the intermediates and transition states involved in these two reactions should be the same. Hence, reductive eliminations can occur through a concerted mechanism, a S_n2 *mechanism*, a radical mechanism or an ionic mechanism. All these mechanisms have been previously described in the section devoted to oxidative addition (Fig. 1.4) and thus, they will be not described again herein.

Concerning the applications of reductive elimination, this reaction is the product-forming step of many catalytic processes [7]. In this thesis, however, since it is focused on the study of Pd-catalyzed C–C cross-coupling reactions, only the role

[10] As in oxidative addition, these trends are only given by way of guidance, because always there may be exceptions.

[11] This principle holds that a reversible reaction proceeds by the same mechanism in both forward and reverse directions.

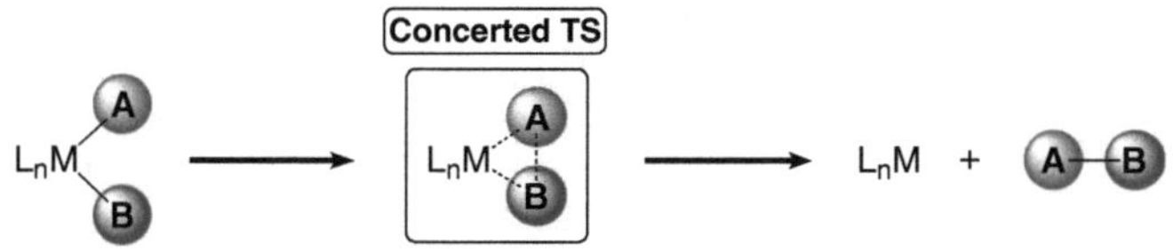

Fig. 1.11 Generally accepted mechanism for the reductive elimination step in cross-coupling reactions

of reductive elimination in these reactions will be discussed. In particular, I will lay stress on some of the most relevant theoretical studies on this step that have been recently reported.

Reductive Elimination in Pd-Catalyzed C–C Cross-Coupling Reactions

The reductive elimination in C–C cross-coupling reactions is the last step of the catalytic cycle and results in the final C–C coupling and the concomitant regeneration of the catalytic species. The generally accepted mechanism for this process is concerted and features a cyclic three-coordinated transition state (Fig. 1.11). Moreover, as reductive elimination is usually irreversible, this step is often taken for granted to be critical for the success of the whole reaction because it must pull the catalytic cycle forward.

Since the early theoretical works of Tatsumi et al. [111] and Low and Goddard [112, 113], the reductive elimination step had received scant attention, but in the last decade, extensive studies on C–C reductive elimination have appeared. In particular, most of these theoretical studies concern the effect of different factors on this step.

In 2002, Ananikov, Musaev and Morokuma reported a theoretical study on the effect of different X ligands on the reductive elimination reaction from bis-σ-vinyl complexes $[Pd(CH{=}CH_2)_2(X)_2]$ ($X = Cl, Br, I, NH_3, PH_3$) revealing that the computed activation barriers decrease in the following order: $Cl > Br \approx NH_3 > I > PH_3$ [114]. A few years later, the same authors investigated most of the common types of coupling partners in the square-planar *cis*-$[Pd(R)(R')(PH_3)_2]$ complexes (R or R' = Me, vinyl, Ph, ethynyl) and found that the Gibbs energy barrier for the carbon-carbon coupling from the symmetrical complex $[Pd(R)_2(PH_3)_2]$ increases in the order: vinyl < Ph < ethynyl < Me [115]. Interestingly, the energy barriers and the exothermicities for the asymmetrical coupling from *cis*-$[Pd(R)(R')(PH_3)_2]$ resulted to be very close to the averages of the corresponding values with the symmetrical R–R and R'–R' coupling reactions from $[Pd(R)_2(PH_3)_2]$ and $[Pd(R')_2(PH_3)_2]$, respectively.

As the other steps in cross-coupling reactions, the rate of the reductive elimination step is also influenced by the steric and electronic effects induced by the ancillary ligands bound to the metal center. Consequently, the influence of the steric and electronic properties of these ligands on this step has been theoretically investigated by several groups. Ananikov, Musaev and Morokuma analyzed by means of the *ONIOM* approach the C–C coupling from the complexes *cis*-$[Pd(R)_2(L)_n]$ (R = Me, Ph, vinyl, ethynyl; L = PPh_3, PCy_3, PMe_3 and PH_3; n = 1, 2) [116]. According

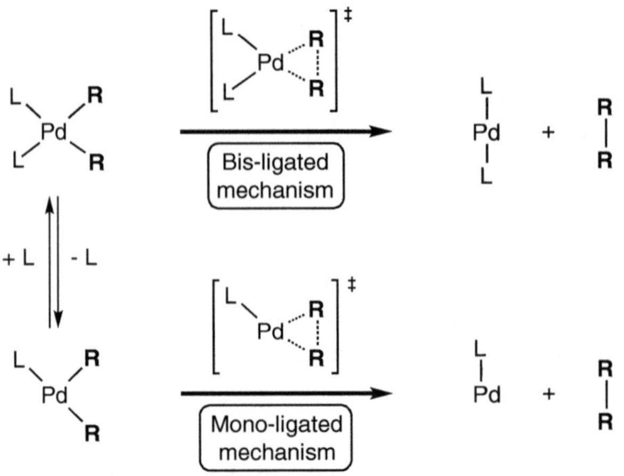

Fig. 1.12 Proposed mechanisms for the reductive elimination reaction from a general *cis*-$[Pd(R)_2(L)_2]$ complex

to the authors, the steric effects mainly influence the energy of the initial complex, whereas the electronic effects have the largest impact on the energy of the transition state. Furthermore, the results revealed that different L ligands may involve different mechanisms of the reductive elimination (Fig. 1.12). Specifically, the PCy_3 ligand was proposed to promote the mono-ligated reaction pathway resulting in an increase in the reactivity, while the PMe_3 ligand was proposed to stabilize the four-coordinated complexes, thus decreasing the reactivity toward reductive elimination. In the particular case of the PPh_3 ligand, this ligand exhibited good reactivity for both mechanisms; hence, the authors referred to this ligand as a "more universal choice".

The influence of the bite angle in chelating bis-phosphine ligands was investigated by Bo et al. revealing that wide bite angle ligands destabilize the four-coordinated intermediate and stabilize the transition state, which reflects in the acceleration of reductive elimination [117].

More recently, the steric and electronic effects of several phosphine ligands ($L = PMe_3$, PH_3, PCl_3, PPh_3, PPh_2Me, $PPhMe_2$) on the Me–Me coupling through the mono-ligated and bis-ligated reaction pathways were investigated by Ariafard and Yates [118]. On the basis of the reported energy barriers, the electronic properties of the ligand (i.e. $L = PMe_3$, PH_3, PCl_3) were found to have a significant influence on the energy of the four-coordinated transition state, whereas its effect on the energy of the three-coordinated transition state was much lower. More specifically, for the bis-ligated mechanism, the authors concluded that the greater the electron-donation of L the higher the energy barrier for this process. In contrast, for the mono-ligated mechanism, the authors proposed the dissociation of the phosphine ligand, which is controlled by the basicity of L, as responsible for the different overall energy barrier; this is, the stronger electron donation leads to a stronger Pd–L bond and, accordingly, to a higher dissociation energy that results in a higher overall barrier.

On the other hand, the energy-decomposition analyses [119, 120] of the reaction barriers with the larger ligands (L = PPh_3, PPh_2Me, and $PPhMe_2$) revealed that the electronic effects are similar in all the series and that the lowering of the energy barrier originates from the steric effects that destabilize the complex *cis*-[$Pd(R)_2(L)_2$] but not the transition state. Thus, the sterically larger phosphines tend to reduce the energy required to dissociate one L ligand, which results in the lowering of the coupling barrier via the mono-ligated mechanism. However, these steric effects do not play a role in the reductive elimination from the three-coordinated complex [$Pd(R)_2(L)$]. In fact, the coupling from this complex was found to be almost independent of the bulk of L. This fact was supported by very small steric repulsions between the L ligand and the Me group calculated in the mono-phosphine species. The reported energy barriers with the larger ligands resulted to increase in the order: $PPh_3 < PPh_2Me < PPhMe_2$ regardless of the mechanism, which agrees with the experimental observations [121].

At the same time that the publication of the work of Ariafard and Yates [118], we reported in collaboration with ?lvarez, Espinet et al. a combined experimental and theoretical study on the C–C coupling from complexes *cis*-[$Pd(R)_2(PMe_3)(L)$] [61]. Specifically, in this study, the mono-phosphine and bis-phosphine pathways (L = "empty" and PMe_3, respectively) with different organic groups (R = Me, Ph, vinyl), as well as the influence on the reaction rate of the addition of several coupling promoters (L = acetonitrile, ethylene, maleic anhydride (MA)), were investigated. As shown in Table 1.1, the trend of computed Gibbs energy barriers at the *DFT-B3LYP* level with different R groups was $Csp^3–Csp^3 > C_{Ar}–C_{Ar} > Csp^2–Csp^2$ for any series with an identical ligand, which agrees with the reported sequence for *cis*-[$Pd(R)_2(PH_3)_2$] complexes [115].

On the other hand, the addition of a coupling promoter led to the following order in the computed energy barriers: L = MA < "empty" < ethylene < $PMe_3 \approx$ MeCN, pointing out that the energy barrier decreases with the π-acceptor ability of L. Thus, the addition of MA consistently produces a lower coupling barrier because it is a better π-acceptor than ethylene, which in turn is better than PMe_3. Furthermore, the energy barrier with L = MA was found to be lower to the point that the coupling with this ligand from the four-coordinated complex resulted to be easier than from the three-coordinated species (L = "empty"). On this basis, we concluded that in cases where a low coupling barrier is operating via a three-coordinated intermediate (e.g. with bulky phosphines), the addition of small electron-deficient olefins with coordinating ability can still help considerably to further accelerate the coupling rate by forming a *cis*-[$Pd(R)_2$(bulky phosphine)(acceptor olefin)] intermediate, from which the coupling takes place with a lower activation energy.

The mono- and bis-phosphine pathways, as well as the effect of some selected ligands (L = "empty", PMe_3, ethylene, maleic anhydride (MA)), were also computationally analyzed very recently by Álvarez, Espinet et al. for the reductive elimination

Table 1.1 Gibbs energy barriers (in kcal·mol^{-1}) for the reductive elimination from *cis*-[Pd(R)$_2$(PMe$_3$)(L)] complexes [61]

R	Ligand L	$\Delta G^\ddagger$
	MA	8.6
	Empty	13.2
Me	CH_2CH_2	21.7
	MeCN	27.0[a]
	PMe_3	28.6
	MA	2.9
	Empty	4.9
Ph	CH_2CH_2	11.3
	MeCN	13.2
	PMe_3	12.8
	MA	0.6
	Empty	4.9
Vinyl	CH_2CH_2	8.9
	MeCN	11.9
	PMe_3	11.5

[a] For this weak ligand a stepwise mechanism consisting in the *MeCN* dissociation followed by the coupling from the resulting three-coordinated complex was predicted. This value corresponds to the overall energy barrier for the reductive elimination reaction through this mechanism

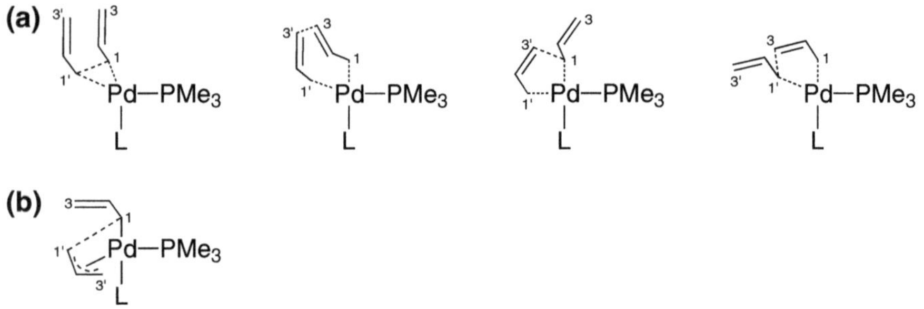

Fig. 1.13 Possible C–C couplings in: a) *cis*-[Pd(η^1-allyl)$_2$(PMe$_3$)(L)] complex; and b) [Pd(η^1-allyl)(η^3-allyl)(L)] complex [122]

reaction from the palladium complexes *cis*-[Pd(η^1-allyl)$_2$(PMe$_3$)(L)] and [Pd(η^1-allyl)(η^3-allyl)(L)] [122]. Unexpectedly, among all the possible carbon-carbon couplings (Fig. 1.13), the C3–C3' bond formation and not the classical C1–C1' was found to be the most favored one in *cis*-[Pd(η^1-allyl)$_2$(PMe$_3$)(L)] complexes, in coincidence with the results previously reported by Echavarren et al. for *cis*-[Pd(η^1-allyl)$_2$(PH$_3$)$_2$] complexes [123]. For this favored coupling, the calculated activation energies with the different ligands increased in the order: "empty" < MA < CH_2CH_2 < PMe_3.

Interestingly, the order in the activation energies of the two most favored C3–C3' allyl couplings (i.e. "empty" < MA) is reversed with respect to the trend found earlier for the (necessary C1–C1') coupling of alkyl, aryl and alkenyl organic groups [61]. This trend, however, reverts to the usual one (i.e. "empty" > MA) in the case of the less favored C1–C1' allyl coupling mechanism. Structural analysis on this particular behavior in the allyl-allyl coupling revealed that the C3–C3' coupling from *cis*-[Pd(η^1-allyl)$_2$(PMe$_3$)] is particularly favored by stereoelectronic effects associated with the small distortion required by the original T-shaped geometry to adopt the transition state geometry while maintaining the stabilizing side-by-side hyper-conjugative d_σ-(C1–Pd)→ π^*(C2=C3) type interaction.

References

1. Berzelius, J.J.: Annales chimie physiques **61**, 146 (1836)
2. Leicester, H.M., Klickstein, H.S.: Source Book in Chemistry 1400–1900. Harvard University Press, Massachusetts (1965)
3. Arrhenius, S.: Zeit. Phys. Chem. **28**, 317 (1899)
4. Rothenberg, G.: Catalysis: Concepts and Green Applications. Wiley-VCH, Weinheim (2008)
5. Eyring, H.: J. Chem. Phys. **3**, 107 (1935)
6. Evans, M.G., Polanyi, M.: Trans. Faraday Soc. **31**, 875 (1935)
7. Hartwig, J.: Organotransition Metal Chemistry: From Bonding to Catalysis. University Science Books, California (2010)
8. Kozuch, S., Shaik, S.: J. Am. Chem. Soc. **128**, 3355 (2006)
9. Kozuch, S., Shaik, S.: J. Phys. Chem. A **112**, 6032 (2008)
10. Uhe, A., Kozuch, S., Shaik, S.: J. Comput. Chem. **32**, 978 (2011)
11. Kozuch, S., Shaik, S.: Acc. Chem. Res. **44**, 101 (2011)
12. Beller, M., Bolm, C. (eds.): Transition Metals for Organic Synthesis, 2nd edn. Wiley-VCH, Weinheim (2004)
13. de Meijere, A., Diedrich, F. (eds.): Metal-Catalyzed Cross-Coupling Reactions, 2nd edn. Wiley-VCH, Weinheim (2004)
14. Miyaura, N. (ed.): Cross-Coupling Reactions: A Practical Guide, No. 219 in Topics in Current Chemistry. Springer, Berlin (2002)
15. Buchwald, S.L. (ed.): Acc. Chem. Res. **41**(special issue), 1439 (2008)
16. Tamao, K., Hiyama, T., Negishi, E.-I.: J. Organomet. Chem. **653**(special issue), 1 (2002)
17. Phapale, V.B., Cárdenas, D.J.: Chem. Soc. Rev. **38**, 1598 (2009)
18. Nicolaou, K.C., Bulger, P.G., Sarlah, D.: Angew. Chem. Int. Ed. **44**, 4442 (2005)
19. Negishi, E.-I., Hu, Q., Huang, Z., Wang, G., Yin, N.: The Chemistry of Organozinc Compounds, pp. 457–553. In: Rappoport, Z.Z., Marek, I. (eds.). Wiley , England (2006)
20. Kosugi, M., Sasazawa, K., Shimizu, Y., Migita, T.: Chem. Lett. **3**, 301 (1977)
21. Stille, J.K.: Angew. Chem. Int. Ed. **25**, 508 (1986)
22. Espinet, P., Echavarren, A.M.: Angew. Chem. Int. Ed. **43**, 4704 (2004)
23. Miyaura, N., Yamada, K., Suzuki, A.: Tetrahedron Lett. **20**, 3437 (1979)
24. Miyaura, N., Suzuki, A.: Chem. Rev. **95**, 2457 (1995)
25. Negishi, E.-I., King, A.O., Okukado, N.: J. Org. Chem. **1977**, 42 (1821)
26. Nicolaou, K. C., Sorensen, E.J.: Classics in Total Synthesis, Chap. 31. VCH, New York (1996)
27. Chemler, S.R., Trauner, D., Danishefsky, S.J., Angew, : Angew. Chem. Int. Ed. **40**, 4544 (2001)
28. Negishi, E.-I.: Bull. Chem. Soc. Jpn. **80**, 233 (2007)
29. Anderson, B., Becke, L., Booher, R., Flaugh, M., Harn, N., Kress, T., Varie, D., Wepsiec, J.: J. Org. Chem. **62**, 8634 (1997)
30. Rouhi, A.M.: Chem. Eng. News **82**, 49 (2004)

31. Zapf, A., Beller, M.: Top. Catal. **19**, 101 (2002)
32. Corbet, J.P., Mignani, G.: Chem. Rev. **106**, 2651 (2006)
33. Zapf, A., Beller, M.: Handbook of Organopalladium Chemistry for Organic Synthesis. In: Negishi, E.-I. (ed.). Wiley, New York (2002)
34. Blaser, H.U., Indolese, A., Naud, F., Nettekoven, U., Schnyder, A.: Adv. Synth. Catal. **346**, 1583 (2004)
35. Tamura, M., Kochi, J.: J. Am. Chem. Soc. **93**, 1483 (1971)
36. Tamura, M., Kochi, J.: J. Am. Chem. Soc. **93**, 1485 (1971)
37. Tamura, M., Kochi, J.K.: J. Organomet. Chem. **42**, 205 (1972)
38. Ishiyama, T., Abe, S., Miyaura, N., Suzuki, A.: Chem. Lett. **691** (1992)
39. Devasagayaraj, A., Studemann, T., Knochel, P.: Angew. Chem. Int. Ed. **34**, 2723 (1996)
40. Giovannini, R., Studemann, T., Dussin, G., Knochel, P.: Angew. Chem. Int. Ed. **37**, 2387 (1998)
41. Giovannini, R., Studemann, T., Devasagayaraj, A., Dussin, G., Knochel, P.: J. Org. Chem. **64**, 3544 (1999)
42. Jensen, A.E., Knochel, P.: J. Org. Chem. **67**, 79 (2002)
43. Frisch, A.C., Beller, M.: Angew. Chem. Int. Ed. **44**, 674 (2005)
44. Littke, A.F., Fu, G.C.: Angew. Chem. Int. Ed. **41**, 4176 (2002)
45. Wolfe, J.P., Singer, R.A., Yang, B.H., Buchwald, S.L.: J. Am. Chem. Soc. **121**, 9550 (1999)
46. Zapf, A., Ehrentraut, A., Beller, M.: Angew. Chem. Int. Ed. **39**, 4153 (2000)
47. Fleckenstein, C.A., Plenio, H.: Chem. Soc. Rev. **39**, 694 (2010)
48. Kantchev, E.A.B., O'Brien, C.J., Organ, M.G.: Angew. Chem. Int. Ed. **46**, 2768 (2007)
49. Marion, N., Nolan, S.P.: Acc. Chem. Res. **41**, 1440 (2008)
50. García-Cuadrado, D., Mendoza, de P., Braga, A.A.C., Maseras, F., Echavarren, A.M.: J. Am. Chem. Soc. **129**, 6880 (2007)
51. Alberico, D., Scott, M.E., Lautens, M.: Chem. Rev. **107**, 174 (2007)
52. Özdemir, I., Demir, S., Çetinkaya, B., Gourlaouen, C., Maseras, F., Bruneau, C., Dixneuf, P.H.: J. Am. Chem. Soc. **130**, 1156 (2008)
53. Ackermann, L., Vicente, R., Kapdi, A.R.: Angew. Chem. Int. Ed. **48**, 9792 (2009)
54. Chiusoli, G.P., Catellani, M., Costa, M., Motti, E., Della Ca', N., Maestri, G.: Coord. Chem. Rev. **254**, 456 (2010)
55. Amatore, C., Jutand, A.: Acc. Chem. Res. **33**, 314 (2000)
56. Galardon, E., Ramdeehul, S., Brown, J.M., Cowley, A., Hii, K.K.M., Jutand, A.: Angew. Chem. Int. Ed. **41**, 1760 (2002)
57. Hartwig, J.F.: Inorg. Chem. **2007**, 46 (1936)
58. Yandulov, D.V., Tran, N.T.: J. Am. Chem. Soc. **129**, 1342 (2007)
59. Senn, H.M., Ziegler, T.: Organometallics **23**, 2980 (2004)
60. Goossen, L.J., Koley, D., Hermann, H.L., Thiel, W.: Organometallics **24**, 2398 (2005)
61. Pérez-Rodríguez, M., Braga, A.A.C., García-Melchor, M., Pérez-Temprano, M.H., Casares, J.A., Ujaque, G., de Lera, A.R., Álvarez, R., Maseras, F., Espinet, P.: J. Am. Chem. Soc. **131**, 3650 (2009)
62. Xue, L., Lin, Z.: Chem. Soc. Rev. **39**, 1692 (2010)
63. Casado, A.L., Espinet, P.: J. Am. Chem. Soc. **120**, 8978 (1998)
64. Goossen, L.J., Koley, D., Hermann, H.L., Thiel, W.: J. Am. Chem. Soc. **127**, 11102 (2005)
65. Sicre, C., Braga, A.A.C., Maseras, F., Cid, M.M.: Tetrahedron **64**, 7437 (2008)
66. Liu, Q., Lan, Y., Liu, J., Li, G., Wu, Y.-D., Lei, A.: J. Am. Chem. Soc. **131**, 10201 (2009)
67. Casares, J.A., Espinet, P., Fuentes, B., Salas, G.: J. Am. Chem. Soc. **129**, 3508 (2007)
68. Álvarez, R., Faza, O.N., López, C.S., de Lera, A.R. Org. Lett. **8**, 35 (2006)
69. Amatore, C., Jutand, A., Le Duc, G.: Chem. Eur. J. **17**, 2492 (2011)
70. Braga, A.A.C., Morgon, N.H., Ujaque, G., Maseras, F.: J. Am. Chem. Soc. **127**, 9298 (2005)
71. Braga, A.A.C., Ujaque, G., Maseras, F.: Organometallics **25**, 3647 (2006)
72. Braga, A.A.C., Morgon, N.H., Ujaque, G., Lledós, A., Maseras, F.: J. Organomet. Chem. **691**, 4459 (2006)

73. Nova, A., Ujaque, G., Maseras, F., Lledós, A., Espinet, P.: J. Am. Chem. Soc. **128**, 14571 (2006)
74. Pérez-Temprano, M. H., Nova, A., Casares, J.A., Espinet, P.: J. Am. Chem. Soc. **130**, 10518 (2008)
75. Fuentes, B., García-Melchor, M., Lledós, A., Maseras, F., Casares, J.A., Ujaque, G., Espinet, P.: Chem. Eur. J. **16**, 8596 (2010)
76. García-Melchor, M., Fuentes, B., Lledós, A., Casares, J.A., Ujaque, G., Espinet, P.: J. Am. Chem. Soc. **133**, 13519 (2011)
77. García-Melchor, M., Pacheco, M.C., Nájera, C., Lledós, A., Ujaque, G.: ACS Catal. **2**, 135 (2012)
78. Noell, J.O., Hay, P.J.: J. Am. Chem. Soc. **104**, 4578 (1982)
79. Low, J.J., Goddard, W.A.: J. Am. Chem. Soc. **106**, 6928 (1984)
80. Obara, S., Kitaura, K., Morokuma, K.: J. Am. Chem. Soc. **106**, 7482 (1984)
81. Bickelhaupt, F.M., Ziegler, T., Schleyer, P.V.R.: Organometallics **14**, 2288 (1995)
82. Ahlquist, M., Norrby, P.O.: Organometallics **26**, 550 (2007)
83. Li, Z., Fu, Y., Guo, Q.-X., Liu, L.: Organometallics **27**, 4043 (2008)
84. Jover, J., Fey, N., Purdie, M., Lloyd-Jones, G.C., Harvey, J.N.: J. Mol. Catal. A **324**, 39 (2010)
85. Schoenebeck, F., Houk, K.N.: J. Am. Chem. Soc. **132**, 2496 (2010)
86. Stille, J.K., Lau, K.S.Y.: Acc. Chem. Res. **10**, 434 (1997)
87. Casado, A.L., Casares, J.A., Espinet, P.: Inorg. Chem. **37**, 4154 (1998)
88. Netherton, M.R., Fu, G.C.: Angew. Chem. Int. Ed. **41**, 3910 (2002)
89. Hills, I.D., Netherton, M.R., Fu, G.C.: Angew. Chem. Int. Ed. **42**, 5749 (2003)
90. Gourlaouen, C., Ujaque, G., Lledós, A., Medio-Simón, M., Asensio, G., Maseras, F.: J. Org. Chem. **74**, 4049 (2009)
91. Crabtree, R.H.: The Organometallic Chemistry of the Transition Metals, 3rd edn. John Wiley-Interscience, New York (2001)
92. Lappert, M., Lednor, P.: Adv. Organomet. Chem. **14**, 345 (1976)
93. Kurosawa, H., Ogoshi, S., Kawasaki, Y., Murai, S., Miyoshi, M., Ikeda, I.: J. Am. Chem. Soc. **112**, 2813 (1990)
94. Kurosawa, H., Kajimaru, H., Ogoshi, S., Yoneda, H., Miki, K., Kasai, N., Murai, S., Ikeda, I.: J. Am. Chem. Soc. **114**, 8417 (1992)
95. Astruc, D.: Organometallic Chemistry and Catalysis. Springer, Berlin (2007)
96. Amatore, C., Jutand, A.: J. Organomet. Chem. **576**, 254 (1999)
97. Kozuch, S., Amatore, C., Jutand, A., Shaik, S.: Organometallics **24**, 2319 (2005)
98. Diefenbach, A., de Jong, G.T., Bickelhaupt, F.M.: J. Chem. Theory Comput. **1**, 286 (2005)
99. Ariafard, A., Lin, Z.: Organometallics **25**, 4030 (2006)
100. Hartwig, J.F., Paul, F.: J. Am. Chem. Soc. **117**, 5373 (1995)
101. Ahlquist, M., Fristrup, P., Tanner, D., Norrby, P.O.: Organometallics **25**, 2066 (2006)
102. Lam, K.C., Marder, T.B., Lin, Z.Y.: Organometallics **26**, 758 (2007)
103. Barrios-Landeros, F., Carrow, B.P., Hartwig, J.F.: J. Am. Chem. Soc. **131**, 8141 (2009)
104. McMullin, C.L., Jover, J., Harvey, J.N., Fey, N.: Dalton Trans. **39**, 10833 (2010)
105. Besora, M., Gourlaouen, C., Yates, B., Maseras, F.: Dalton Trans. **40**, 11089 (2011)
106. García-Melchor, M., Ujaque, G., Maseras, F., Lledós, A., in Phosphorus Compounds: Advanced Tools in Catalysis and Material Sciences, vol. 37., Chap. 3, pp. 57–84 . In: Peruzzini, M., Gonsalvi, L. (eds.). Springer, Berlin (2011)
107. Farina, V.: Comprehensive Organometallic Chemistry II, vol. 12, Chap. 3.4. In: Abel, E.W., Stone, F., Wilkinson, G., Hegedus, L.S. Elsevier, Oxford (1995)
108. Osakada, K.: Fundamentals of Molecular Catalysis, Current Methods in Inorganic Chemistry, Chap. 5. In: Kurosawa, H., Yamamoto, A. (eds.). Elsevier, Amsterdam (2003)
109. Marshall, J.A.: Chem. Rev. **100**, 3163 (2000)
110. Pérez-Temprano, M.H., Gallego, A.M., Casares, J.A., Espinet, P. Organometallics, **30**, 611 (2011)
111. Tatsumi, K., Hoffmann, R., Yamamoto, A., Stille, J.K.: Bull. Chem. Soc. Jpn. **1981**, 54 (1857)
112. Low, J.J., Goddard, W.A.: J. Am. Chem. Soc. **108**, 6115 (1986)

113. Low, J.J., Goddard, W.A.: Organometallics **5**, 609 (1986)
114. Ananikov, V.P., Musaev, D.G., Morokuma, K.: J. Am. Chem. Soc. **124**, 2839 (2002)
115. Ananikov, V.P., Musaev, D.G., Morokuma, K.: Organometallics **24**, 715 (2005)
116. Ananikov, V.P., Musaev, D.G., Morokuma, K.: Eur. J. Inorg. Chem. 5390 (2007)
117. Zuidema, E., van Leeuwen, P.W.N.M., Bo, C.: Organometallics **24**, 3703 (2005)
118. Ariafard, A., Yates, B.F.: J. Organomet. Chem. **694**, 2075 (2009)
119. Ziegler, T., Rauk, A.: Theor. Chim. Acta. **46**, 1 (1977)
120. Bickelhaupt, F.M.: J. Comput. Chem. **20**, 114 (1999)
121. Negishi, E.-I., Takahashi, T., Akiyoshi, K.: J. Organomet. Chem. **334**, 181 (1987)
122. Pérez-Rodríguez, M., Braga, A.A.C., de Lera, A.R., Maseras, F., Álvarez, R., Espinet, P. Organometallics, **29**, 4983 (2010)
123. Méndez, M., Cuerva, J.M., Gómez-Bengoa, E., Cárdenas, D.J., Echavarren, A.M.: Chem. Eur. J. **8**, 3620 (2002)

Chapter 2
Computational Methods

We are perhaps not far removed from the time when we shall be able to submit the bulk of chemical phenomena to calculation.
Joseph Louie Gay-Lussac (1888)

The aim of this chapter is not just to make a general description of the computational methods used in this thesis (as it is usually the case), but also to try to provide an overview of what quantum mechanics is, and how do we use it to study chemical reactions. In particular, the first part of this chapter will be devoted to make a brief introduction on the origin and some basic concepts of quantum mechanics. In general, this is a part that is usually skipped by the Ph.D. candidates probably because they are not experts on it, or just because collecting the main ideas of quantum mechanics in a reasonable space is a harsh work. Honestly, I would probably be in the first group. However, when I started "doing bibliography" for this chapter, I found something that changed my mind. Specifically, it was a sentence from one of most brilliant physicists, Richard Feynman, who said: *"I think I can safely say that nobody understands quantum mechanics"*. Obviously, this sentence comforted me and was what prompted me to write and include this first part of the chapter. In any case, I am aware that there may be people that may know much more about quantum mechanics than I do, and that might not agree with some aspects of this first part of the chapter. To those, I sincerely apologize.

In the second part of this chapter, I will explain how we, computational chemists, apply quantum mechanics to the study of chemical reactions, which are the approximations that we do, and how do we obtain valuable information by means of applying these approximations.

Finally, in the last part of this chapter, I will briefly describe the methods that I have used during this thesis. This last part is probably the part that involves more mathematical equations, but don't worry about it because I will only include the most important ones. Furthermore, my intention in this last part of the chapter is not to do an exhaustive description of the methods, but to provide a comprehensive view of the original concepts, approximations and ideas that are behind them.

M. G. Melchor, *A Theoretical Study of Pd-Catalyzed C–C Cross-Coupling Reactions*, Springer Theses, DOI: 10.1007/978-3-319-01490-6_2,

2.1 Quantum Mechanics Basis

2.1.1 Origin and Basic Concepts

As we go deeper into the microscopic world, classical mechanics laws start failing and do not provide the good results that they are used to. This impossibility of reproducing some observations by means of classical mechanics is a consequence of the so-called *quantum realm*, which is a term applied in Physics that refers to the scales where quantum mechanical effects become important.[1] This fact, in particular, was what led to the development of quantum mechanics in the first half of 20th century. However, it is important not to forget that, despite the differences between classical and quantum mechanics, they are not independent. In fact, according to the *correspondence principle* between classical and quantum mechanics, all objects obey the laws of quantum mechanics, and classical mechanics is just an approximation for large systems (or a statistical quantum mechanics of a large collection of particles). Thus, classical and quantum mechanics provide the same answer when the systems under study become large. These conditions under which quantum and classical mechanics agree are commonly known as the *correspondence limit* or *classical limit*.

One of the basic concepts derived from quantum mechanics is the *discretization of magnitudes* (e.g. energy), which states that magnitudes can only have discrete values or *quantized* values. Other derived basic concepts are the *wave-particle duality*, which postulates that all particles exhibit both wave and particle properties, and the *uncertainty principle*. This last principle, formulated by Werner Heisenberg in 1927, [1] states a fundamental limit on the accuracy with which certain pairs of physical properties of a particle, such as position and momentum, can be simultaneously known. In other words, the more precisely one property is measured, the less precisely the other can be determined. Thus, quantum mechanics only provides a range of probabilities of where a particle might be given its momentum and momentum probability. This range of probabilities are described by the *wavefunction* (Ψ) and depends on the quantum state at the "instant" of the measurement. Hence, uncertainty is indeed involved in the value. Nevertheless, there are certain states that are associated with a definite value of a particular measured property (i.e. *observable*); these states are the *eigenstates* of the observable. Unfortunately, usually a system is not in an eigenstate of the observable that we are interested in. However, if we measure the observable, the wavefunction will instantaneously be an eigenstate of that observable. This process, known as *wavefunction collapse*, has been a much debated process but due to the lack of space, time, and deeper knowledge, it will not be further discussed in this thesis.

[1] Interestingly, these scales are typically distances of 100 nm or less, which is (and not coincidentally) the same scale as Nanotechnology.

2.1.2 The Schrödinger Equation

With a role similar to *Newton's second law* in classical mechanics, the *Schrödinger equation* in quantum mechanics describes how the wavefunction evolves in time:

$$i\hbar\frac{\partial}{\partial t}\Psi(x,t) = H\Psi(x,t) \tag{2.1}$$

In this last equation, H is the *Hamiltonian*, an operator that generates the time evolution of quantum states and provides the value for the total energy of the system. By analogy with classical mechanics, this operator is generally expressed as the sum of the operators for the kinetic (T) and potential (V) energies[2]:

$$H = T + V \tag{2.2}$$

As with all the operators, there are a series of states that are eigenstates of the Hamiltonian, whose eigenvalues correspond to the energies of those states. In fact, if we assume that the Hamiltonian does not contain time,[3] the spatial and time dependences of the wavefunction for these states can be treated separately:

$$\Psi(x,t) = f(t)\Psi(x) \tag{2.3}$$

Then, if the wavefunction $\Psi(x)$ is an eigenstate of the Hamiltonian, the Eq. (2.1) can be written as:

$$i\hbar\frac{1}{f(t)}\frac{\partial f(t)}{\partial t} = \frac{1}{\Psi(x)}H\Psi(x) = E \tag{2.4}$$

where the constant E that appears in this equation is the total energy of the system in the quantum state described by the wavefunction.

Now, if we isolate the two function parts of Eq. (2.4) we obtain the following two equations:

$$f(t) = e^{-iEt/\hbar} \tag{2.5}$$

and

$$H\Psi(x) = E\Psi(x) \tag{2.6}$$

This last Eq. (2.6) is usually called *time-independent Schrödinger* equation to distinguish it from Eq. (2.1), commonly named *time-dependent Schrödinger* equation.

[2] In this general definition of the Hamiltonian, relativistic effects are neglected. These effects are normally negligible for the first three rows in the periodic table (i.e. $Z < 36$), but become important for the fourth and fifth rows, and for transition metals. Other operators such as the ones describing spin-orbit, orbit-orbit, or spin-spin couplings are also neglected because their contributions are, in most cases, rather small.

[3] This implies that the total energy of the system remains constant; thus, the system is conservative.

It is interesting to note that Eq. (2.6) can also be seen as the spatial part of the time-dependent Schrödinger equation (2.1), because the time dependency is always an exponential function of the type represented in the Eq. (2.5).

Hence, the eigenstates of the Hamiltonian are of the type:

$$\Psi(x,t) = e^{-iEt/\hbar}\,\Psi(x) \tag{2.7}$$

Importantly, these eigenstates have the peculiarity that they produce probability distributions that are independent of time and, accordingly, their properties do not depend on it either. For this reason, they are called *stationary states*. Among these stationary states, the one with the lowest energy value is commonly known as the *ground state* and the other ones as *excited states*.

In general, most of the methods used for the study of chemical reactions focus on the resolution of the time-independent Schrödinger equation (2.6). Thus, from now on, we will refer to this last equation simply as "*Schrödinger equation*", unless otherwise specified.

2.2 Theory Applied to Reactivity

After having summarized the origin and some of the basic concepts of quantum mechanics in the previous section, the next step is to understand how quantum mechanics is applied to the study of chemical reactions.

The application of quantum mechanics to the study of reactivity results in very complex equations, which have led to the appearance of multiple approximations. These approximations allow obtaining approximate values for many properties of molecules. However, their acceptance without having checked their validity can occasionally lead to cases where they are not valid and, consequently, to incorrect results. On the other hand, the no acceptance of approximations, even when they are valid, restricts the study to only very small systems though, in such cases, we have the certainty of obtaining the correct results. Hence, when facing chemical problems, the choice (or not) of these approximations is crucial.

This section, in particular, will be devoted to summarize which are the common approximations that we use when applying quantum mechanics for studying chemical reactions, and how do we obtain valuable information of these reactions by means of these approximations. It should be said, however, that this section does not intend to be exhaustive on the theoretical methods to study reactivity, but just to introduce how do we approach to such analysis.

2.2.1 The Born-Oppenheimer Approximation and the Concept of Electronic Structure

The *Born-Oppenheimer* approximation is probably the approximation that has been most successfully applied in Theoretical Chemistry. In general terms, we could say that it allows the decomposition of the Schrödinger equation in two parts: one part that describes the electronic wavefunction for a fixed nuclear geometry, and the other part that describes the nuclear wavefunction, where the energy from the electronic wavefunction plays the role of a potential energy. This is, however, a very generalized description of this approximation, so let us go into a more detailed description of it.

The Born-Oppenheimer approximation is based on the following two assumptions:

- *The wavefunction can be separated by the product of an electronic wavefunction and a nuclear wavefunction:*

$$\Psi_{tot}(R, r) = \Psi_{el}(R, r)\ \Psi_{nuc}(R) \tag{2.8}$$

  (R and r denote the nuclear and the electronic coordinates, respectively).
- *The electronic wavefunction depends parametrically*[4] *on the coordinates of the nuclei, but not on their momenta.*

The basis for these two assumptions is the high mass ratio between nuclei and electrons, which reflects in a much faster movement of the electrons compared to the nuclei. In general, we can safely say that this is a very good approximation because for the lightest nuclei, the *proton* or ^1H, that ratio is: $\frac{m_{^1H}}{m_{e^-}} \approx 1836$.

The Born-Oppenheimer approximation consists of two steps that we will discuss below. First, let us start by writing the Schrödinger equation for a given molecule:

$$H_{tot}\Psi_{tot}(R, r) = E_{tot}\Psi_{tot}(R, r) \tag{2.9}$$

The total Hamiltonian, H_{tot}, in Eq. (2.9) is:

$$H_{tot} = T_{tot} + V_{tot} = (T_{el} + T_{nuc}) + (V_{ne} + V_{ee} + V_{nn}) \tag{2.10}$$

where T_{el} and T_{nuc} stand for the kinetic energy of electrons and nuclei, respectively, V_{ne} stands for the coulombic attraction between electrons and nuclei, and V_{ee} and V_{nn} stand for the coulombic repulsion between electrons and between nuclei, respectively.

Now, assuming that nuclei move much more slowly than electrons, and that the latter are moving in the potential of fixed nuclei, it results that the term corresponding to the nuclear kinetic energy can be dropped from Eq. (2.10) and that the repulsion between the nuclei can be considered to be constant. At this point, we can define the *electronic Hamiltonian* as:

---

[4] By *parametrically* we mean that for different arrangements of the nuclei, the electronic wavefunction is a different function of the electronic coordinates.

$$H_{el} = T_{el} + (V_{ne} + V_{ee} + \underbrace{V_{nn}}_{const}) = T_{el} + V \tag{2.11}$$

Then, if we apply this electronic Hamiltonian to the electronic wavefunction, we obtain the electronic state of the molecule by solving the equation:

$$(T_{el} + V)\Psi_{el}(R, r) = U_n(R)\Psi_{el}(R, r) \tag{2.12}$$

where U_n is the sum of the electronic energy and the potential energy of the nuclei, for a fixed nuclear coordinates. This dependence of U_n on the nuclear coordinates means that for each nuclear positions, we will have a different electronic wavefunction and a different electronic energetic spectrum. The resolution of this Eq. (2.12) is the first of the two steps in the Born-Oppenheimer approximation.

Once the electronic Schrödinger equation (2.12) has been solved, the second step is to solve the complete Schrödinger equation. For this step, we just have to apply the complete Hamiltonian (2.10) to the nuclear wavefunction:

$$(T_{nuc} + T_{el} + V)\Psi_{nuc}(R) = E_{tot}\Psi_{nuc}(R) \tag{2.13}$$

Then, as the electrons move much faster than the nuclei, it is a reasonable approximation to replace the electronic Hamiltonian by its average value. This generates a nuclear Hamiltonian for the motion of the nuclei in the average potential of the electrons, which transforms Eq. (2.13) into:

$$(T_{nuc} + U_n(R))\Psi_{nuc}(R) = E_{tot}\Psi_{nuc}(R) \tag{2.14}$$

This potential U_n, also known as *adiabatic potential*, comes from the resolution of the electronic Schrödinger equation (2.12) and, as above mentioned, corresponds to the sum of the electronic energy and the coulombic repulsion between the nuclei, for a fixed nuclear arrangement. This adiabatic potential calculated for a large number of nuclear geometries is known as *potential energy surface* (*PES*).

Overall, in the Born-Oppenheimer approximation, the nuclei move on a potential energy surface obtained by solving the electronic Schrödinger equation. This assumption of the nuclei moving on a potential energy surface was, in particular, what led to the development of methods aimed at solving the electronic Schrödinger equation. These methods are broadly referred to as *electronic structure calculations*, and allow obtaining molecular properties that result very useful in the study of chemical reactions.

2.2.2 *The Potential Energy Surface (*PES*)*

As previously stated, the potential energy surface (*PES*) is nothing else than the terms U_n that one obtain in the Born-Oppenheimer approximation by solving the

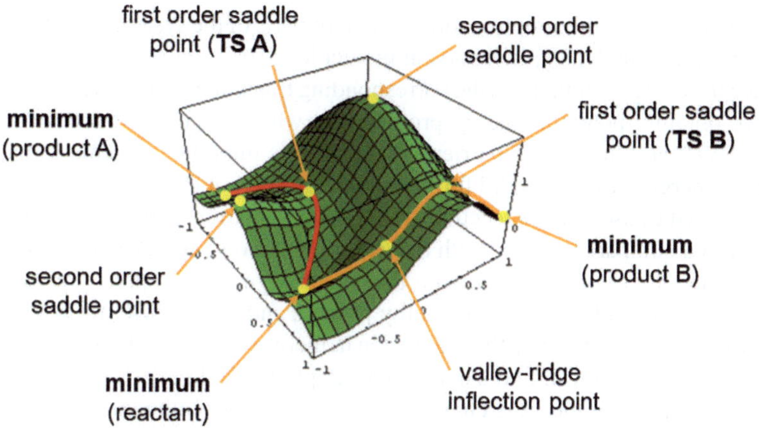

Fig. 2.1 Example of a *3D*-representation of an arbitrary *PES*

electronic Schrödinger equation (2.12) for a set of fixed nuclear coordinates. Hence, once the electronic Schrödinger equation has been solved for all the possible nuclear configurations, the complete *PES* is known. Unfortunately, the construction of the complete *PES* for molecules containing more than 3–4 atoms is virtually impossible given that it is a $3N$-6 dimension space, where N is the number of atoms.

Hence, a general strategy in computational chemistry consists in restricting the calculations to the chemically interesting part of the *PES*, which usually allows extracting useful information. The most interesting points from the chemical perspective are, in general, the points on the *PES* where the energy is stationary with respect to the nuclear coordinates. Particularly, those that are minima and first-order saddle points (Fig. 2.1).

The energy minima give us an idea of the expected nuclear configuration of a molecular structure, and are potentially observables as stable products or reaction intermediates. On the other hand, the first-order saddle points (or *transition states, TSs*) are points on the *PES* that are minima in all directions but one, which are maxima. Furthermore, the energy of these transition states provide an estimation of the reaction rate when going from one minimum to another one. Thus, the reaction minima and the transition states that relate these minima are usually the stationary points on the *PES* that we look for when we are interested in the reaction mechanism and/or the kinetics of a particular reaction. Finding these stationary points, however, is frequently not trivial[5] and requires the choice of an appropriate methodology, which entails a right balance between accuracy and computational cost. This choice of the methodology includes the selection of the level of theory (e.g. Molecular Mechanics, Quantum Mechanics, *QM/MM*), the method (e.g. Force Field, Hartree-Fock, *DFT*, *MP2*), the basis sets (e.g. double-ζ, pseudo-potentials, polarization functions), the

[5] Some of us even think that the location of transition states can be considered as an "art".

chemical model (i.e. the simplification, or not, of parts of the system), and the solvent (e.g. continuum model, explicit solvent molecules).

Once the energy minima and the corresponding transition states have been located, then we obtain a potential energy profile from which we can extract the reaction energy (ΔE) and the energy barrier(s) ($\Delta E^{\ddagger}$). It is important to remark that these potential energies are only valid for one molecule and not for an ensemble of molecules, as it is the case in a real reaction. Thus, if we want to calculate a thermodynamic property and compare its value with the experimental one, we have to use statistical thermodynamics in order to estimate this thermodynamic property for the whole ensemble of molecules. This is a lengthy process and thus, it will not be discussed here. However, if you have further interest in it, there are excellent books on statistical mechanics where you can find a detailed description of it [2–4].

As we will see in the next chapters, one of the most common thermodynamic properties that we compute for its further comparison with the experimental value is the *Gibbs energy* in solution (G_s).[6] Generally, this thermodynamic magnitude is estimated by computing the entropic contributions for the solute in gas phase (i.e. in vacuum). This, of course, is a rough approximation, but normally provides good results except for associative and dissociative processes. In these last cases, the results through this approximation are not so good mainly because, in gas phase, the translational and rotational contributions to the entropy are overestimated. This overestimation stems from the fact that, in solution, the molecules are surrounded by the solvent and consequently, they can not freely move and rotate as they do in gas phase. Hence, the computation of dissociative processes in condensed phases, such as the dissociation of one molecule into two fragments, results to be more favorable than what they really are. On the contrary, the computation of associative processes turn out to be less favorable than what they are. On this issue, several approaches have been developed and proposed in the last decades [5, 6]. For example, some authors [7–9] have proposed to only consider the vibrational contribution to the entropy, but if the previous approximation overestimates the entropy, this one underestimates it. Another approximation is the one proposed by Ziegler et al. [10–12] and based on the *Wertz model* [13]. This approximation, however, requires the knowledge of reference quantities, which sometimes might be not available.[7] On the other hand, Martin et al. [14] proposed simulating the condensed phase by computing the Gibbs energies in gas phase at high pressure. More specifically, the authors suggested that the computation of the Gibbs energies at the pressure: $p = \rho_W RT$, is a good estimation.[8] This is also a rough approximation because it involves the modification of the standard state

---

[6] When we talk about "experimental Gibbs energies", we should be aware that these are not truly experimental values, since there is no *"Gibbs-meter"* that enables measuring this thermodynamic property. In fact, these experimental Gibbs energies are always based on theoretical models. Thus, even though we use the term "experimental Gibbs energies", they are as theoretical as the ones that we compute.

[7] This approximation in the case of solvation entropies in water states that they can be approximated in qualitative discussions as 50 % of the gas phase entropy, with the opposite sign.

[8] In this formula, ρ_W is the experimental density of the solvent, R is the ideal gas constant, and T is the temperature.

(T = 298.15 K, and p = 1 atm) but, somehow, it allows simulating the constraints imposed by the solvent over the free movement of the molecules.

Obviously, apart from all these approximations, there are other methods that, in principle, provide better Gibbs energy values (e.g. molecular dynamics simulations, Monte-Carlo-based methods). These methods, however, are too computationally demanding and thus, they are normally restricted to small systems.

2.2.3 Approximate Resolution of the Schrödinger Equation

In the case of one-electron systems, like in the H_2^+ molecule, the electronic Schrödinger equation (2.12) can be solved exactly; however, this is not the case when we are in front of many-electron systems. Typically, we say that this impossibility of solving exactly this equation is because the Schrödinger equation is a *many-body interacting* problem. But, what does it exactly mean? Well, this means that, as a consequence of the electron-electron repulsive interactions, we can not solve the problem of N-interacting electrons as a sum of N one-electron problems. In order to overcome this issue, several methods have been developed, which, by means of different approximations, deal with this electronic problem. These methods aimed at solving the electronic Schrödinger equation can be classified in two well differentiated classes: the methods that are based on the wavefunction, and the methods that are based on the electron density. Among the first class of methods, probably the most important one is the *Hartree-Fock* (*HF*) method, since most of the precise methods are based on corrections to this method. As far as the methods based on the electron density is concerned, these methods have received special attention and are being increasingly used due to the good results that they provide at a reasonable computational cost.

In this thesis, all the calculations have been performed using methods based on the electron density. Thus, the following section will be exclusively devoted to describe these methods, and the approximations that there are behind them.

2.3 The Density Functional Theory (DFT)

In the methods based on the wavefunction, the resolution of the electronic Schrödinger equation (2.12) implies the computation of the wavefunction for a given nuclear coordinates, which is a function of 3N spatial coordinates plus N spin variables, where N is the number of electrons. This, in the systems that we are usually interested in, where there are many atoms and many more electrons, becomes a very complicated problem. But, is there any other formulation that allows simplifying this problem? The answer is *yes*. This alternative formulation is the so-called *Density Functional Theory* (*DFT*) and is based on the calculation of electronic properties from the electron density, instead of from the wavefunction. This formulation has the main advantage that the electron density depends only on the three spatial coordinates (*x*, *y*, *z*), which simplifies considerably the problem.

The electron density $\rho(\vec{r})$ is defined as the multiple integral over the spin variables of all the electrons and over all but one of the spatial coordinates,

$$\rho(\vec{r}_1) = N \int \cdots \int |\Psi(\vec{x}_1, \vec{x}_2, \ldots, \vec{x}_N)|^2 ds_1 d\vec{x}_2 \ldots d\vec{x}_N \qquad (2.15)$$

where

$$d\vec{x}_i = ds_i d\vec{r}_i$$

The electron density represents the probability of finding any of the N electrons within the volume element $d\vec{r}_1$ but with arbitrary spin, while the other N-1 electrons have arbitrary positions and spin in the state represented by Ψ. Thus, strictly speaking $\rho(\vec{r})$ is a probability density although, in practice, it is commonly known simply as electron density. Furthermore, since the electrons are indistinguishable, the probability of finding any electron at that position is just N times the probability for one particular electron. Therefore, we can say that $\rho(\vec{r})$ is a non-negative function of only three spatial coordinates that vanishes at infinity and integrates to the total number of electrons:

$$\rho(\vec{r} \to \infty) = 0, \qquad (2.16)$$

$$\int \rho(\vec{r}_1) d\vec{r}_1 = N \qquad (2.17)$$

2.3.1 The Hohenberg-Kohn Theorems

The density functional theory as we know it today was born in 1964 with the appearance of the seminal paper of Hohenberg and Kohn [15]. In this paper, the authors stated two theorems that now represent the major theoretical bedrock of *DFT*. Quoting directly from the Hohenberg and Kohn's paper, the first theorem states:

"The external potential $V_{ext}(\vec{r})$ is (to within a constant) a unique functional of $\rho(\vec{r})$; since, in turn, $V_{ext}(\vec{r})$ fixes H we see that the full many-particle ground state is a unique functional of $\rho(\vec{r})$".

The authors demonstrated this first theorem by proving by *reductio ad absurdum* that there can not be two different $V_{ext}(\vec{r})$[9] that result in the same ground state electron density; or what is the same, the ground state density uniquely specifies the external potential $V_{ext}(\vec{r})$. A direct consequence derived from this first principle is that all the ground state properties of a system are defined by its electron density.

[9] This external potential is the potential that the nuclei exert over the electrons.

On the other hand, given that the complete ground state energy is a functional[10] of the ground state electron density so must be its individual components. Thus, the expectation value for the energy of the system can be written as[11]:

$$E[\rho] = T[\rho] + V_{ee}[\rho] + V_{ne}[\rho] \tag{2.18}$$

Now, if we separate the terms that are dependent on the external potential and those which are *universal* or independent on it, we arrive at

$$E[\rho] = \underbrace{T[\rho] + V_{ee}[\rho]}_{independent} + \underbrace{V_{ne}[\rho]}_{dependent} = F_{HK}[\rho] + \int \rho_r V_{ne} d\vec{r} \tag{2.19}$$

where the independent terms have been collected into a new quantity, the *Hohenberg-Kohn functional*, $F_{HK}[\rho]$:

$$F_{HK}[\rho] \equiv T[\rho] + V_{ee}[\rho] \tag{2.20}$$

According to the Eq. (2.19), if we know $F_{HK}[\rho]$, then we can solve the Schrödinger equation exactly. Moreover, it is noteworthy to note that since this is a functional that is completely independent of the system, it applies equally well to the hydrogen atom as to huge molecules such as the *DNA*! Unfortunately, things usually are not as easy as they seem, and this case will not be an exception. Here, the main problem is that the exact form of the functional $F_{HK}[\rho]$ is not known. Hence, the pursuit of the explicit expression for this unknown functional is the major challenge in the *DFT* and, at the same time, represents the weakest point of this theory.

Up to this point, we have seen that the ground state density is, in principle, sufficient to obtain all the properties of interest. However, how can we know that a certain density is indeed the ground state density that we are looking for? Well, here is where the second Hohenberg-Kohn theorem takes part. This second theorem states:

The functional $F_{HK}[\rho]$, which provides the ground state energy of the system, provides the lowest energy if and only if the input density is truly the ground state density.

This statement is nothing else than applying the variational principle established for wavefunctions, to the present case:

$$E_0 \leq E[\tilde{\rho}] = T[\tilde{\rho}] + V_{ee}[\tilde{\rho}] + V_{ne}[\tilde{\rho}] \tag{2.21}$$

In this last equation, the density $\tilde{\rho}(\vec{r})$ represents a trial density that defines a trial external potential, a trial Hamiltonian, and a trial wavefunction, and leads to an energy that is higher than or equal to the exact energy of the system. Thus, in order

[10] A functional is a prescription for producing a number from a function, which in turn depends on variables. Thus, the wavefunction and the electron density are functions, whereas the energy that depends on the wavefunction or the electron density, is a functional.

[11] In this expression, the Born-Oppenheimer approximation has been adopted.

to obtain the exact energy of the system we will have to find the electron density that minimizes the energy,

$$\left[\frac{\partial E[\rho]}{\partial \rho}\right] = 0 \tag{2.22}$$

Overall, the second Hohenberg-Kohn theorem establishes the variational principle in the *DFT*. However, we have to be aware of what it implies and do not overinterpret this result. Many conventional methods based on the wavefunction, such as the *HF* method, are strictly variational and, consequently, the expectation value of the energy is an indicator of the quality of the trial wavefunction. This means that the lower the energy the better approximated the trial wavefunction to the ground state wavefunction. However, this is not the case in the "real" *DFT*.[12] The variational principle stated by the second Hohenberg-Kohn theorem applies only for the exact functional, which is not known. Thus, the energy provided by a trial functional has absolutely no meaning in that respect.

2.3.2 The Kohn-Sham Method

As above mentioned, the ignorance of the exact expression of the functional $F_{HK}[\rho]$ is undoubtedly the major drawback when trying to solve a system of N electrons by means of the *DFT*. An alternative to overcome this issue is to use an approximate expression for this functional and solve the N-electron problem approximately. In fact, if the approximate functional $F_{HK}[\rho]$ is good enough, we should obtain an energy value close to the exact energy of the system. This aim at obtaining an approximate expression for $F_{HK}[\rho]$ was what led to the formulation of the *Thomas-Fermi* (*TF*) and *Thomas-Fermi-Dirac* (*TFD*) methods [16–18]. These two methods are normally viewed as precursors of modern *DFT*, and further represent the basis of the formulation of the widely known *Kohn-Sham* (*KS*) method [19]. In particular, this last method is considered as the second major paper of the modern *DFT*, and is based on corrections to the *TF* and *TFD* methods. Thus, let us first start by making a brief summary of these two methods before going into the description of the Kohn-Sham method.

In the *TF* and *TFD* methods, the general expression for the energy functional (2.18),

$$E[\rho] = T[\rho] + V_{ee}[\rho] + V_{ne}[\rho]$$

is approximated as follows: the first term $T[\rho]$, which corresponds to the kinetic energy of the N-electron system, is substituted by the functional corresponding to a non-interacting and uniform electron gas, $T_{TF}[\rho]$. On the other hand, the term that stands for the electron-electron repulsion, $V_{ee}[\rho]$, is replaced by the classic

[12] When I say "real" *DFT* I am assuming that the exact expression of the functional $F_{HK}[\rho]$ is not known.

coulombic repulsion between two electrons, $J[\rho]$, and the exchange energy, $K[\rho]$.[13] Taking into account all these approximations, the equation for the energy functional $E[\rho]$ in the *TFD* method can be written as:

$$E_{TFD}[\rho] = T_{TF}[\rho] + J[\rho] + K[\rho] + V_{ne}[\rho] \tag{2.23}$$

With the expression for the energy functional already defined, let us see now how do the terms in (2.23) look like. In the *TF* and *TFD* methods, the terms $V_{ne}[\rho]$ and $J[\rho]$ are treated in a completely classical way, so they can be easily obtained from their classical expressions,

$$V_{ne}[\rho] = -\sum_{a}^{M} \int \frac{Z_a \rho(\vec{r})}{|R_a - \vec{r}|} d\vec{r} \tag{2.24}$$

$$J[\rho] = \frac{1}{2} \iint \frac{\rho(\vec{r_1})\rho(\vec{r_2})}{|\vec{r_1} - \vec{r_2}|} d\vec{r_1} d\vec{r_2} \tag{2.25}$$

where the factor of 1/2 in $J[\rho]$ allows the integration to run over all space for both variables.

As regards to the terms $T_{TF}[\rho]$ and $K[\rho]$, it can be demonstrated that their expressions for a non-interacting uniform gas of electrons are:

$$T_{TF}[\rho] = C_F \int \rho^{5/3}(\vec{r}) d\vec{r} \tag{2.26}$$

$$K[\rho] = -C_x \int \rho^{4/3}(\vec{r}) d\vec{r} \tag{2.27}$$

with

$$C_F = \frac{3}{10}(3\pi^2)^{2/3} = 2.8712 \tag{2.28}$$

$$C_x = \frac{3}{4}\left(\frac{3}{\pi}\right)^{1/3} = 0.7386 \tag{2.29}$$

Hence, introducing the expressions for $T_{TF}[\rho]$, $J[\rho]$, $K[\rho]$, and $V_{ne}[\rho]$ into the equation for the energy functional $E[\rho]$ (2.23), we obtain that the expression for the energy functional in the *TFD* method is given by:

[13] This exchange energy term is only added in the *TFD* method.

$$E_{TFD}[\rho] = \underbrace{C_F \int \rho^{5/3}(\vec{r})d\vec{r}}_{T_{TF}[\rho]} + \underbrace{\frac{1}{2}\iint \frac{\rho(\vec{r_1})\rho(\vec{r_2})}{|\vec{r_1}-\vec{r_2}|}d\vec{r_1}d\vec{r_2}}_{J[\rho]}$$
$$- \underbrace{C_x \int \rho^{4/3}(\vec{r})d\vec{r}}_{K[\rho]} - \underbrace{\sum_a^M \int \frac{Z_a\rho(\vec{r})}{|R_a-\vec{r}|}d\vec{r}}_{V_{ne}[\rho]} \quad (2.30)$$

This expression (2.30) but without the term $K[\rho]$ corresponds to the energy functional in the *TF* method.

To summarize, with the *TF* and *TFD* methods all the terms included in the energy functional (2.30) are explicit functionals of the electron density, which simplifies considerably the problem of the N-electron system. However, these two methods fail miserably in many cases, mainly because the approximation that these models do for the kinetic energy term is rather simple. Here was when, in 1965, Kohn and Sham [19] realized that if we are not able to accurately determine the kinetic energy through an explicit functional of the density, we should concentrate on calculating as much as we can the true kinetic energy exactly. Thus, their basic idea was to obtain from a Slater determinant[14] the exact kinetic energy of a non-interacting reference system with the same density as the real one (i.e. the interacting system),

$$T_S = -\frac{1}{2}\sum_i^N \langle\varphi_i|\nabla^2|\varphi_i\rangle \quad (2.31)$$

Of course, the kinetic energy of this non-interacting system is not equal to the kinetic energy in the real system, even if both systems share the same density. In order to account for this and the other approximations present in the universal functional $F_{HK}[\rho]$ (Eq. 2.20), Kohn and Sham collected all the necessary corrections into a new term E_{XC}, so that the expression for $F_{HK}[\rho]$ becomes:

$$F[\rho] = T_S[\rho] + J[\rho] + E_{XC}[\rho] \quad (2.32)$$

This term $E_{XC}[\rho]$ is known as *exchange-correlation energy*, and is defined as:

$$E_{XC}[\rho] \equiv (T[\rho] - T_S[\rho]) + (V_{ee}[\rho] - J[\rho]) \quad (2.33)$$

The two first terms in $E_{XC}[\rho]$ correct the difference between the true and the approximate kinetic energies, while the last two terms correct the difference between the classical electron-electron interaction $J[\rho]$ defined in (2.25) and the real one.

[14] A Slater determinant is an antisymmetrized product of N one-electron wavefunctions (*spin orbitals*). In this determinant, the columns are the one-electron wave functions while the electron coordinates are along the rows.

Thus, we can say that the exchange-correlation energy $E_{XC}[\rho]$ is the functional that contains everything that is unknown.

Introducing the expression (2.32) into (2.19), we can write down the expression for the energy functional of the real system as:

$$\begin{aligned} E[\rho(\vec{r})] &= T_S[\rho] + J[\rho] + E_{XC}[\rho] + V_{ne}[\rho] \\ &= -\frac{1}{2}\sum_i^N \langle\varphi_i|\nabla^2|\varphi_i\rangle \quad + \quad \frac{1}{2}\iint \frac{\rho(\vec{r_1})\rho(\vec{r_2})}{|\vec{r_1}-\vec{r_2}|} d\vec{r_1} d\vec{r_2} \\ &+ E_{XC}[\rho] \quad - \quad \sum_a^M \int \frac{Z_a \rho(\vec{r})}{|R_a - \vec{r}|} d\vec{r} \end{aligned} \tag{2.34}$$

Now, introducing into this last equation the restriction that the electron density resulting from the summation of the moduli of the squared orbitals φ_i exactly equals the ground state density of our real system of interacting electrons,

$$\rho(\vec{r}) = \sum_i^N \sum_s |\varphi_i(\vec{r}, s)|^2 = \rho_0(\vec{r}) \tag{2.35}$$

and imposing the constraint that the orbitals φ_i have to be orthonormal,

$$\begin{aligned} \int \varphi_i(\vec{r})^* \varphi_j(\vec{r}) &= \delta_{ij} \\ \delta_{ij} &= 1 \quad i = j \\ \delta_{ij} &= 0 \quad i \neq j \end{aligned} \tag{2.36}$$

we finally arrive to the Kohn-Sham equations:

$$\underbrace{\left(-\frac{1}{2}\nabla^2 + v_{eff}\right)}_{h_{KS}} \varphi_i = \varepsilon_i \varphi_i \tag{2.37}$$

where h_{KS} is the Kohn-Sham monoelectronic hamiltonian and the term v_{eff} is the *effective potential*, which is defined as:

$$v_{eff}(\vec{r}) = \int \frac{\rho(\vec{r_2})}{r_{12}} d\vec{r_2} + V_{XC}(\vec{r_1}) - \sum_a^M \frac{Z_a}{r_{1a}} \tag{2.38}$$

In this last equation, the only term that is unknown is the potential V_{XC} due to the exchange-correlation energy E_{XC}. This potential is simply defined as the functional derivative of E_{XC} with respect to the electron density:

$$V_{XC} \equiv \frac{\partial E_{XC}}{\partial \rho} \tag{2.39}$$

It is very important to realize that if the exact forms of E_{XC} and V_{XC} were known, the Kohn-Sham method would provide the exact energy. Unfortunately, this is not the case. Furthermore, since the effective potential depends on the electron density, the Kohn-Sham Eqs. (2.37) have to be solved in an iterative way. For this process, we define a trial electron density from which we can calculate the effective potential through Eq. (2.38). Then, with this effective potential we solve the Kohn-Sham Eqs. (2.37) and obtain the orbitals φ_i, which are introduced into the Eq. (2.35) resulting in a new electron density. This process is iteratively repeated until the difference between this new electron density and the trial density satisfies the desired convergence criterion. Once this is done, the energy can be easily computed from Eq. (2.34) using the converged electron density.

Another important point that should be remarked is that unlike the *HF* method, where the approximation is introduced right from the start,[15] the Kohn-Sham method is in principle exact! The approximation in this last method only comes into play when we choose an explicit form of the unknown exchange-correlation functional E_{XC} and the corresponding potential V_{XC}. Therefore, the difference between *DFT* methods stem from the form of these unknown functionals.

2.3.3 Exchange-Correlation Functionals

The form of exchange-correlation functionals is often designed to have a certain limiting behavior (e.g. including the uniform electron gas limit), and fitting parameters to known accurate data. A common approach in the design of these functionals consists in splitting them into two parts: the exchange part E_X, and the correlation part E_C

$$E_{XC}[\rho] = E_X[\rho] + E_C[\rho] \tag{2.40}$$

These exchange and correlation parts are normally calculated separately by means of different approximations and then, combined into the exchange-correlation functional. Based on these approximations, John P. Perdew presented by analogy with the biblical referent of *Jacob's ladder* [20], his particular vision of which are the rungs to be climbed in the *DFT* for going from the Earth (i.e. the *HF* world) to Heaven (i.e. the chemical accuracy) (Fig. 2.2).

According to Perdew, the main advantage of the *DFT* is that we can go up or down this ladder of functionals depending on our necessities. In principle, going up this ladder involves an improvement of the results but also, an increase in the computational cost. Thus, when choosing an exchange-correlation functional, we

---

[15] In the *HF* method the wavefunction is assumed to be a single Slater determinant, which therefore, can never provide the true solution.

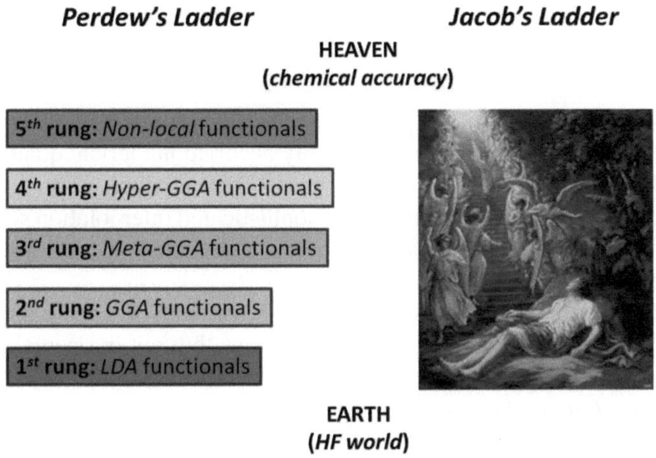

Fig. 2.2 Graphical representation of the "Perdew's functionals ladder" (*left*) and the Jacob's Ladder (*right*)

should arrive to a good compromise between accuracy and computational cost. It is important, however, to keep in mind that despite of this general classification of functionals, the typical one million dollar question *"which is the best functional?"* does not have a unique answer, and needs to be settled for each particular case by comparing the performance of functionals with experiments or high-level wave mechanics calculations.

In the next pages, I will briefly describe the different types of exchange-correlation functionals that appear in the Perdew's ladder (Fig. 2.2), and which are the approximations that are behind them.

Local Density and Local Spin Density Approximations (LDA/LSDA)

The *local density approximation* (*LDA*) constitutes the simplest way of representing the exchange-correlation functional. In particular, this approximation assumes that the density can be locally treated as a uniform electron gas. The general expression for the functional E_{XC} in the *LDA* is:

$$E_{XC}^{LDA}[\rho] = \int f(\rho) d\vec{r} = E_X^{LDA}[\rho] + E_C^{LDA}[\rho] \tag{2.41}$$

The exchange part in a uniform electron gas is nothing else that the exchange functional that we met in the Thomas-Fermi-Dirac method (2.27). Thus, the exchange energy E_X in this *LDA* approximation is also given by:

$$E_X^{LDA}[\rho] = -\frac{3}{4}\left(\frac{3}{\pi}\right)^{1/3} \int \rho^{4/3}(\vec{r})d\vec{r} \tag{2.42}$$

As far as the correlation part E_C^{LDA} is concerned, no such explicit expression for this term is known. However, there are highly accurate numerical quantum Monte-Carlo simulations of the uniform electron gas [21] from which several authors have derived analytical expressions by means of sophisticated interpolation schemes. One of the most widely used representations for this term is the one developed by Vosko, Wilk, and Nusair (*VWN*) [22].

In the *LDA*, the total density is considered to be the sum of α and β spin densities. This assumption is satisfied in closed-shell systems, but not in open-shell systems. If we extend the *LDA* to the latter case, we arrive to the so-called *local spin density approximation* (*LSDA*). The exchange energy in this *LSDA* is:

$$E_X^{LSDA}[\rho] = -(2)^{1/3}\ \frac{3}{4}\left(\frac{3}{\pi}\right)^{1/3} \int [\rho_\alpha^{4/3} + \rho_\beta^{4/3}]d\vec{r} \tag{2.43}$$

Despite the simplicity of these *LDA* or *LSDA* approximations, it has been proven that they can deliver results with a similar accuracy than those obtained with the *HF* method. However, we should not get overexcited about these results because a look at energetical properties, such as bond energies, immediately shows that the performance of these approximations is rather poor. Hence, there is still a long way to go in the development of exchange-correlation functionals before the chemical accuracy can be reached. The next step on this way is the introduction of the gradient of the density, which takes us to the next rung of the Perdew's ladder (Fig. 2.2): the *generalized gradient approximation* (*GGA*) functionals.

The Generalized Gradient Approximation (GGA)

Probably the most logical and intuitive way of improving the *LDA*-based functionals is to try to account for the non-homogeneity of the true electron density. A first approximation to this aim is to consider that exchange and correlation energies not only depend on the electron density, but also on its derivatives. This, in particular, represents the basis of the *generalized gradient approximation* (*GGA*) functionals.[16] The general expression for the exchange-correlation functional in the *GGA* is:

$$E_{XC}^{GGA}[\rho] = \int f(\rho, \nabla\rho)d\vec{r} = E_X^{GGA}[\rho] + E_C^{GGA}[\rho] \tag{2.44}$$

---

[16] These functionals are sometimes referred to as non-local functionals in order to distinguish them from the *LDA* ones. However, it is more appropriate to consider them as local functionals because they depend only on the density (and derivatives) at a given point, and not on a space volume as the *HF* exchange energy.

The exchange part in this approximation can be expressed as:

$$E_X^{GGA}[\rho] = E_X^{LDA}[\rho] - \sum_{\sigma} \int F(s_\sigma)\rho_\sigma^{4/3}(\vec{r})d\vec{r} \quad (2.45)$$

where F is a function whose argument is the *reduced density gradient* for spin σ,[17]

$$s_\sigma(\vec{r}) = \frac{|\nabla \rho_\sigma(\vec{r})|}{\rho_\sigma^{4/3}(\vec{r})} \quad (2.46)$$

In general, most of the *GGA* exchange functionals are developed following two main philosophies. The first one, initiated with the formulation of the *GGA* exchange functional *B* (or *B88*) by Becke [23], is based on fitting empirical parameters usually considering experimental data for a set of atoms or model molecules. Some examples of *GGA* exchange functionals that follow this philosophy are the *PW91* [24–27], the *OPTX* (*O*) [28, 29], and the modified Perdew-Wang (*mPW*) functional [30]. On the other hand, the second philosophy considers that the development of exchange functionals should be based on principles derived from quantum mechanics. Examples of *GGA* exchange functionals following this second philosophy are the *B86* functional from Becke [31], the *P* functional from Perdew [32], and the *PBE* functional from Perdew, Burke, and Ernzerhof [33].

Regarding the *GGA* correlation functionals, these have an even more complicated analytical form than the exchange functionals. Thus, I will merely enumerate some of the most commonly used correlation functionals. For example, it is worth highlighting the correlation counterpart of the exchange functional *P* from Perdew (*P* or *P86*) [32], the *LYP* functional from Lee, Yang and Parr [34], and the parameter-free correlation functional *PW91* from Perdew and Wang [24–27]. In principle, all the correlation functionals could be combined with any of the exchange functionals, but in practice, only some combinations are currently in use.

Meta-GGA Functionals

According to Perdew's ladder (Fig. 2.2), the next rung corresponds to the *meta-GGA* functionals [35]. The difference between this third rung and the previous one (i.e. the *GGA* rung) is that this type of exchange-correlation functionals expands GGA to include further the kinetic energy density τ (i.e. the Laplacian of the occupied orbitals) and/or the Laplacian of the density $\nabla^2\rho(r)$.

$$\tau(\vec{r}) = \frac{1}{2} \sum_{i}^{occupied} |\nabla \varphi_i(\vec{r})|^2 \quad (2.47)$$

[17] The term s_σ can be understood as a local inhomogeneity parameter.

Thus, the *meta-GGA* functionals are functionals of the density, the gradient of the density, and the kinetic energy density.

$$E_{XC}^{meta-GGA}[\rho] = \int f(\rho, \nabla\rho, \tau) d\vec{r} \tag{2.48}$$

Some examples of this type of functionals are the exchange-correlation functionals *B98* [36], *TPSS* [37], *VSXC* [38] and the correlation functional *KCIS* [39–42].

The Adiabatic Connection: Hybrid Functionals

Keeping up with the quest for a most accurate exchange-correlation functional, another appropriate strategy is to use the exact exchange energy calculated with the *HF* method and the Kohn-Sham orbitals, and rely on approximate functionals only for the part that *HF* misses, i.e. the electron correlation (2.49). The exchange-correlation functionals based on this strategy are commonly known as *hybrid functionals*.

$$E_{XC} = E_X^{exact} + E_C^{KS} \tag{2.49}$$

The connection between these two components of the exchange-correlation energy is achieved by means of the so-called *adiabatic connection formula* (*ACF*),

$$E_{XC}[\rho] = \int_0^1 E_{ncl}^{\lambda}[\rho] d\lambda \tag{2.50}$$

where λ is the *coupling strength parameter* (with values between 0 and 1), and E_{ncl} is nothing else than the non-classical contribution to the electron-electron interaction for different values of λ. In particular, at the integration limit $\lambda = 0$, we are dealing with a non-interacting system and accordingly, $E_{ncl}^{\lambda=0}$ is composed of exchange only. Hence, this limiting case (i.e. $\lambda = 0$) simply corresponds to the exact exchange. On the other hand, at $\lambda = 1$, the non-classical contributions are those of the fully interacting system, which contains both exchange and correlation terms. This interacting exchange-correlation is not known and therefore, has to be approximated by any E_{XC} functional.

The simplest approximation to solve the Eq. (2.50) is to assume that E_{ncl}^{λ} is a linear function in λ, and use the *LDA* exchange-correlation functional for $E_{ncl}^{\lambda=1}$. This approximation leads to the so-called *half-and-half* (*HH*) combination proposed by Becke [43], in which a 50 % of exact exchange and a 50 % of *LDA* exchange is included.

$$E_{XC}^{HH} = \frac{1}{2} E_X^{\lambda=0} + \frac{1}{2} E_X^{\lambda=1} \tag{2.51}$$

One of the most widely used variants of this combination is the *BHandHLYP* [23, 34, 43], which has been shown to deliver particularly good results for radical systems [44–47].

$$E_{XC}^{BHandHLYP} = \frac{1}{2}E_X^{exact} + \frac{1}{2}E_X^{LSDA} + \frac{1}{2}\Delta E_X^{B88} + E_C^{LYP} \quad (2.52)$$

During the same year than the proposal of the *HH* combination and following the same scheme, Becke developed one of the most famous hybrid functionals, i.e. the *Becke 3 parameter functional* (*B3*) [48]. For this functional, the expression for the exchange-correlation is given by:

$$E_{XC}^{B3} = aE_X^{exact} + (1-a)E_X^{LSDA} + b\Delta E_X^{B88} + E_C^{LSDA} + c\Delta E_C^{PW91} \quad (2.53)$$

where *a*, *b*, and *c* are parameters that weigh the various terms in this *B3* functional. In particular, these three parameters were chosen to optimally reproduce a set of properties such as total energies, atomization and ionization energies, and proton affinities. The resulting values for these parameters are: $a = 0.20$, $b = 0.72$, and $c = 0.81$. Note that in this functional, the parameter *a* determines the amount of exact exchange (i.e. 20 %).

The substitution of the correlation terms in this *B3* functional (2.53) by the correlation functional *LYP* was proposed by Stevens et al. [49], and gave birth to the appearance of the widely known *B3LYP* functional.[18]

$$E_{XC}^{B3LYP} = aE_X^{exact} + (1-a)E_X^{LSDA} + b\Delta E_X^{B88} + cE_C^{LYP} + (1-c)E_C^{VWN} \quad (2.54)$$

Since its appearance in 1994, the *B3LYP* functional has experienced an unprecedented success to the point that, in the last years, it has been by far the most popular and most widely used functional. This hegemony, however, is presently getting lost mainly because *B3LYP* (as most of the functionals) fails in certain aspects, such as the bad description of non-covalent interactions (e.g. van der Waals interactions) and the underestimation of energy barriers [50, 51]. Thus, new functionals and/or correction terms have been developed to overcome these lacks. Among these new functionals one could highlight the appearance of the *"MXX"* series of functionals of Truhlar et al. which have shown to provide satisfactory results. The first of these series of functionals are the *M05* functionals [52–54], which consist of two functionals: the *M05* [52] and the *M05-2X* [53]. In particular, the *M05* functional (with 28 % of exact exchange) is recommended by the authors for reactivity studies involving metals, while the *M05-2X* functional (with 56 % of exact exchange) is recommended for reactivity studies without metals. The second of these series are the *M06* functionals [54–57], which includes four functionals: the *M06* functional [55] (with 27 % of exact exchange), the *M06-2X* functional [55] (with 54 % of exact exchange), the *M06-L* functional [56] (without exact exchange), and the *M06-HF* functional [57] (with 100 % of exact exchange). At present, the last of theses series of functionals are the *M08* functionals, which includes the *M08-HX* functional (with 52.23 % of

[18] In this functional, since the *LYP* functional contains a local part besides of the correction to the gradient, the term E_C^{VWN} is also added in order to subtract this local part. This last term corresponds to the *LSDA*-based correlation functional derived from Vosko, Wilk, and Nusair [22].

exact exchange) and the *M08-SO* functional (with 56.79 % of exact exchange) [58], and the *M11* functional [59].

Unlike the *B3LYP* functional, where the non-covalent interactions are not well described, most of Truhlar's *"MXX"* functionals are designed in order to provide a good description of these interactions. This, in particular, is carried out by means of implicit parametrization of these functionals. However, there are other ways of improving the description of non-covalent interactions in *DFT* [60–63]. Among them, probably the most popular and in fashion strategy up to date is the method of dispersion correction (*DFT-D*), commonly known as *Grimme's dispersion correction* [60]. This correction consists in the addition to the Kohn-Sham energy of a dispersion energy term defined as:

$$E_{disp} = -s_6 \sum_{i=1}^{N_{at}-1} \sum_{j=i+1}^{N_{at}} \frac{C_6^{ij}}{R_{ij}^6} f_{dmp}(R_{ij}) \tag{2.55}$$

where N_{at} is the number of atoms in the system, C_6^{ij} denotes the dispersion coefficient for atom pair ij, s_6 is a global scaling factor that only depends on the density functional used, f_{dmp} is a damping function, and R_{ij} is an interatomic distance. Recently, some additional corrections to this dispersion term have been added resulting in a considerably improvement of the performance of these dispersion corrected density functionals [64–66].

Hyper-GGA Functionals

Going another rung up in the Jacob's ladder of functionals (Fig. 2.2) we arrive to the level of the *hyper-GGA* functionals. Some books and articles define the *hyper-GGA* functionals as functionals that include exact exchange. This definition, however, is rather incomplete because it does not specify whether a functional that includes part of exact exchange (e.g. hybrid functionals) can be considered as a *hyper-GGA* functional or not. On the other hand, there are many articles and books that consider that the term *hyper-GGA* applies to functionals that include part of exact exchange. Thus, according to this last definition, hybrid functionals can be considered as *hyper-GGA* functionals. Honestly, I have to recognize that even after having surfed the literature looking for the exact meaning of this *hyper-GGA* term, its exact definition remains still not clear for me. In fact, in my humble opinion, its general definition entails certain ambiguity. In any case, this is just a mere question of nomenclature and has no further relevance, but it is interesting to see how this term can lead to two different concepts.

Importantly, even though the *hyper-GGA* functionals appear as the last rung in the Perdew's ladder (Fig. 2.2), there is still another rung before reaching the chemical accuracy. This last level corresponds to fully *non-local functionals*, which includes the exact exchange and refines the correlation part by evaluating part of it exactly.

Examples of this last type of functionals are the *generalized random phase approximation* (*RPA*) [67–72] and the *interaction strength interpolation* (*ISI*) [73].

2.3.4 Self-Interaction Error

Apart from the problem of the ignorance of the exact form of the exchange-correlation functional, there is one more typical problem in the *DFT* methods: the so-called *self-interaction error*. Probably the easiest way of understanding this error is to consider the particular case of a one-electron system and compare the energies obtained with the Kohn-Sham method and the *HF* method, which is free of self-interaction errors. Let us first start by the *HF* method.[19]

In the *HF* method, the electronic energy of the system is given by:

$$E_{HF} = \sum_{i=1}^{N} \varepsilon_i - \frac{1}{2}\sum_{i=1}^{N}\sum_{j=1}^{N}(J_{ij} - K_{ij}) \tag{2.56}$$

where ε_i is the energy of the electron i, and J and K are the coulombic and exchange integrals, respectively. In a monoelectronic system, since there is no electron-electron repulsion, the terms J and K in (2.56) are equal with opposite sign, which leads to the complete cancellation of the self-interaction of the electron i.

On the other hand, in the Kohn-Sham scheme (2.34),

$$E[\rho(\vec{r})] = T_S[\rho] + J[\rho] + E_{XC}[\rho] + V_{ne}[\rho]$$

the term $J[\rho]$

$$J[\rho] = \frac{1}{2}\iint \frac{\rho(\vec{r_1})\rho(\vec{r_2})}{|\vec{r_1} - \vec{r_2}|} d\vec{r_1} d\vec{r_2}$$

is also different from zero because it contains the spurious interaction of the density with itself, but in this case, does not cancel completely with E_{XC}. This no complete cancellation stems from the approximate expressions we have to do for the unknown term E_{XC}. A solution to this problem is the *self-interaction corrected* (*SIC*) form of approximate functionals suggested by Perdew and Zunger [77].

[19] For the lack of time and space, the expression for the electronic energy in the *HF* method has not been derived in this thesis. Even so, I will use this expression in order to explain the concept of self-interaction error. Details on the derivation of this equation can be found in many books on general quantum chemistry [74–76].

2.4 Computation of Solvent Effects

In general, most of the chemical reactions are carried out in the presence of a solvent, which can, in some cases, be crucial for the final outcome of the reaction (e.g. stabilizing species, acting as a reactant). Thus, when studying chemical reactions, an appropriate description of solvent effects is highly recommended. But, how do we introduce solvent effects into calculations? One alternative for evaluating solvent effects is to consider individual solvent molecules explicitly. This alternative, however, involves dealing with a higher number of molecules, which increases the number of degrees of freedom and therefore, the computational cost. Furthermore, other problems that can arise related to this alternative are, for example, which is the number of molecules that should be included, and where should they be placed.[20] Hence, this alternative is normally used only in specific cases where the solvent actively participates in the reaction, and provided they were not too computationally demanding.

A more affordable alternative is to treat the solvent as a continuous medium with the commonly named *continuum methods*. Among this type of methods, probably the most widely used is the *Self-Consistent Reaction Field* (*SCRF*) [78], which considers the solvent as a uniform polarizable medium with a dielectric constant ε, and with the solute placed in a suitable shaped hole in the medium. In this method, the electric charge distribution of the solute polarizes the medium, which in turn acts back on the solute, thereby producing an electrostatic stabilization. This process is iteratively repeated until the mutual polarization between the solute and solvent achieves the self-consistency.[21]

The solvation *"free"* energy[22] in the *SCRF* method can be calculated as a sum of different energy contributions:

$$\Delta G_{solvation} = \Delta G_{cav} + \Delta G_{disp} + \Delta G_{rep} + \Delta G_{elec} \tag{2.57}$$

In this expression, the ΔG_{cav} term corresponds to the energetic cost that is needed to create the cavity for the solute and thus, this term is always positive. On the other hand, the ΔG_{disp} term is associated with the dispersive interactions between solute and solvent, and contributes positively to the solvation. In contrast, the ΔG_{rep} term is associated with the repulsive interactions between solute and solvent, and is always positive. Finally, the term ΔG_{elec} is a stabilizing term that accounts for the electrostatic interactions between the solute and solvent.

Within the *SCRF* method, we can distinguish between different models depending on several features:

---

[20] To overcome this issue one should use methods such as molecular dynamics or Monte-Carlo-based methods.

[21] This is the reason for the name of this method.

[22] Even though it is generally considered as a Gibbs energy, it is really not, because entropic terms are not included.

- How the size and shape of the cavity is defined.
- How the dispersion contributions are calculated.
- How the charge distribution of the solute is represented.
- How the solute is described (i.e. classically or quantally).
- How the dielectric constant is described.

In this thesis, the models that have been used for all the calculations are the widely known *polarizable continuum model* (*PCM*) [79], and the recently developed *SMD* model [80]. These two models define the cavity for the solute as the union of a series of interlocking spheres centered on the atoms and differ only in that the latter includes the radii and non-electrostatic terms as suggested by Truhlar and coworkers. Other variants of the *PCM* model are, for example, the *Isoelectronic-PCM* (*IPCM*), which uses a static isodensity surface for the cavity, and its improved version *self-consistent isodensity-PCM* (*SCI-PCM*) [81].

In all *PCM* methods, the radii of the spheres used to create the cavity can be further defined in different ways, but in the course of this thesis only the two most common radii have been employed: the *UA0* and *UFF*. These two radii use the united atom topological model applied on atomic radii of the *UFF* force field for heavy atoms and only differ in how the spheres for hydrogen atoms are described. More specifically, in the *UA0* radii the hydrogen atoms are enclosed in the sphere of the heavy atom to which they are bonded, while in the *UFF* radii they have individual spheres.

In a compromise between good description of solvent effects and computational cost, combinations of discrete (i.e. with explicit solvent molecules) and continuum methods are a good alternative. An example of combination of these methods is to consider explicitly the first solvation sphere, and treat the rest with a continuous method. This combination is commonly referred to as *discrete-continuum* methods and has been used in some parts of the present thesis.

References

1. Heisenberg, W.: A hadrons and nuclei. Zeitschrift für Physik **43**, 172 (1927)
2. Gasser, R.P.H., Richards, W.G.: An Introduction to Statistical Thermodynamics. World Scientific Publishing, London (1995)
3. McQuarrie, D.A.: Statistical Mechanics. University Science Books, California (2000)
4. Atkins, P., de Paula, J.: Physical Chemistry, 9th ed., vol. 2. W. H. Freeman & Co., New York (2010)
5. Zhou, H.-X., Gilson, M.K.: Chem. Rev. **109**, 4092 (2009)
6. Spickermann, C.: Entropies of Condensed Phases and Complex Systems: A First Principles Approach. Springer, Berlin (2011)
7. Tamura, H., Yamazaki, H., Sato, H., Sakaki, S.: J. Am. Chem. Soc. **125**, 16114 (2003)
8. Sakaki, S., Takayama, T., Sumimoto, M., Sugimoto, M.: J. Am. Chem. Soc. **126**, 3332 (2004)
9. Sumimoto, M., Iwane, N., Takahama, T., Sakaki, S.: J. Am. Chem. Soc. **126**, 10457 (2004)
10. Cooper, J., Ziegler, T.: Inorg. Chem. **41**, 6614 (2002)
11. Zhu, H., Ziegler, T.J.: J. Organomet. Chem. **691**, 4486 (2006)
12. Zhu, H., Ziegler, T.: Organometallics **27**, 1743 (2008)
13. Wertz, D.H.: J. Am. Chem. Soc. **102**, 5316 (1980)

14. Martin, R.L., Hay, P.J., Pratt, L.R.: J. Phys. Chem. A **102**, 3565 (1998)
15. Hohenberg, P., Kohn, W.: Phys. Rev. B **136**, 864 (1964)
16. Thomas, L.H.: Proc. Camb. Phil. Soc. **23**, 542 (1927)
17. Fermi, E.: Rend. Accad. Lincei. **6**, 602 (1927)
18. Dirac, P.A.M.: Proc. Camb. Phil. Soc. **26**, 376 (1930)
19. Kohn, W., Sham, L.: J. Phys. Rev. A **140**, 1133 (1965)
20. Genesis, 28:11–19
21. Ceperley, D.M., Alder, B.: J. Phys. Rev. Lett. **45**, 566 (1980)
22. Vosko, S.J., Wilk, L., Nusair, M.: Can. J. Phys. **58**, 1200 (1980)
23. Becke, A.D.: Phys. Rev. A **38**, 3098 (1988)
24. Perdew, J.P.: Electronic Structure of Solids. In: Ziesche, P., Eschrig, H. (eds.) Akademie Verlag, Berlin (1991)
25. Perdew, J.P., Chevary, J.A., Vosko, S.H., Jackson, K.A., Pederson, M.R., Singh, D.J., Fiolhais, C.: Phys. Rev. B. **46**, 6671, erratum Phys. Rev. B **1993**(48), 4978 (1992)
26. Perdew, J.P., Burke, K., Wang, Y.: Phys. Rev. B **54**, 16533 (1996)
27. Burke, K., Perdew, J.P., Wang, Y.: Electronic Density Functional Theory. Recent Progress and New Directions. In: Dobson, J.F., Vignale, G., Das, M.P. (eds.) Plenum Press, New York (1998)
28. Handy, N.C., Cohen, A.J.: Mol. Phys. **99**, 403 (2001)
29. Hoe, W.-M., Cohen, A., Handy, N.C.: Chem. Phys. Lett. **341**, 319 (2001)
30. Adamo, C., Barone, V.: J. Chem. Phys. **108**, 664 (1998)
31. Becke, A.D.: J. Chem. Phys. **84**, 4524 (1986)
32. Perdew, J.P.: Phys. Rev. B **33**, 8822 (1986)
33. Perdew, J.P., Burke, K., Ernzerhof, M.: Phys. Rev. Lett. **77**, 3865, erratum Phys. Rev. Lett. **1997**(78), 1396 (1996)
34. Lee, C., Yang, W., Parr, R.G.: Phys. Rev. B **37**, 785 (1988)
35. Perdew, J.P., Kurth, S., Zupan, A., Blaha, P.: Phys. Rev. Lett. **82**, 2544 (1999)
36. Schmider, H.L., Becke, A.D.: J. Chem. Phys. **108**, 9624 (1998)
37. Tao, J.M., Perdew, J.P., Staroverov, V.N., Scuseria, G.E.: Phys. Rev. Lett. **91**, 146401 (2003)
38. Van Voorhis, T., Scuseria, G.E.: J. Chem. Phys. **109**, 400 (1998)
39. Rey, J., Savin, A.: Int. J. Quantum Chem. **69**, 581 (1998)
40. Krieger, J.B., Chen, J.Q., Iafrate, G.J., Savin, A.: Electron Correlations and Materials Properties, pp. 463–477. In: Gonis, A., Kioussis, N., Ciftan, M. (eds.) Kluwer Academic, New York (1999)
41. Krieger, J.B., Chen, J.Q., Kurth, S.: Density Functional Theory and its Application to Materials, vol. 577, pp. 48–69. In: van Doren, V., van Alsenoy, C., Geerlings, P. (eds.) A.I.P., New York (2001)
42. Toulouse, J., Savin, A., Adamo, C.: J. Chem. Phys. **117**, 10465 (2002)
43. Becke, A.D.: J. Chem. Phys. **98**, 1372 (1993)
44. Poater, J., Solà, M., Rimola, A., Rodríguez-Santiago, L., Sodupe, M.: J. Phys. Chem. A **108**, 6072 (2004)
45. Georgieva, I., Trendafilova, N., Rodríguez-Santiago, L., Sodupe, M.: J. Phys. Chem. A **109**, 5668 (2005)
46. Rimola, A., Rodríguez-Santiago, L., Sodupe, M.: J. Phys. Chem. B **110**, 24189 (2006)
47. Rimola, A., Constantino, E., Rodríguez-Santiago, L., Sodupe, M.: J. Phys. Chem. A **112**, 3444 (2008)
48. Becke, A.D.: J. Chem. Phys. **98**, 5648 (1993)
49. Stephens, P.J., Devlin, F.J., Chabalowski, C.F., Frisch, M.J.: J. Phys. Chem. **98**, 11623 (1994)
50. Gonzales, J.M., Cox III, R.S., Schaefer III, H.F.: J. Phys. Chem. A **105**, 11327 (2001)
51. Zhao, Y., Truhlar, D.G.: Acc. Chem. Res. **41**, 157 (2007)
52. Zhao, Y., Schultz, N.E., Truhlar, D.G.: J. Chem. Phys. **123**, 161103 (2005)
53. Zhao, Y., Schultz, N.E., Truhlar, D.G.: J. Chem. Theory Comput. **2**, 364 (2006)
54. Zhao, Y., Truhlar, D.G.: Acc. Chem. Res. **41**, 157 (2008)
55. Zhao, Y., Truhlar, D.G.: Theor. Chem. Acc. **120**, 215 (2008)
56. Zhao, Y., Truhlar, D.G.: J. Chem. Phys. **125**, 194101 (2006)
57. Zhao, Y., Truhlar, D.G.: J. Phys. Chem. A **110**, 13126 (2006)

58. Zhao, Y., Truhlar, D.G.: J. Chem. Theory Comput. **2008**, 4 (1849)
59. Peverati, R., Zhao, Y., Truhlar, D.G.: J. Phys. Chem. Lett. **2**, 2810 (2011)
60. Grimme, S.: J. Comput. Chem. **25**, 1463 (2004)
61. Sato, T., Tsuneda, T., Hirao, K.: Mol. Phys. **103**, 1151 (2005)
62. Von Lilienfeld, O.A., Tavernelli, I., Röthlisberger, U., Sebastiani, D.: Phys. Rev. Lett. **93**, 153004 (2004)
63. Sun, Y.Y., Kim, Y.-H., Lee, K., Zhang, S.B.: J. Chem. Phys. **129**, 154102 (2008)
64. Grimme, S.: J. Comput. Chem. **27**, 1787 (2006)
65. Grimme, S., Antony, J., Ehrlich, S., Krieg, H.: J. Chem. Phys. **132**, 154104 (2010)
66. Grimme, S., Ehrlich, S., Goerigk, L.: J. Comput. Chem. **32**, 1456 (2011)
67. Langreth, D.C., Perdew, J.P.: Phys. Rev. B **21**, 5469 (1980)
68. Pitarke, J.M., Eguiluz, A.G.: Phys. Rev. B **63**, 045116 (2001)
69. Furche, F.: Phys. Rev. B **64**, 195120 (2001)
70. Fuchs, M., Gonze, X.: Phys. Rev. B **65**, 235109 (2002)
71. Yan, Z., Perdew, J.P., Kurth, S.: Phys. Rev. B **61**, 16430 (2000)
72. Lein, M., Gross, E.K.U., Perdew, J.P.: Phys. Rev. B **61**, 13431 (2000)
73. Seidl, M., Perdew, J.P., Kurth, S.: Phys. Rev. Lett. **84**, 5070 (2000)
74. Szabo, A., Ostlund, N.S.: Modern Quantum Chemistry. Dover Publications, INC., New York (1996)
75. Jensen, F.: Introduction to Computational Chemistry. John Wiley & Sons, England (1999)
76. Andrés, J., Bertran, J.: Química Teórica y Computacional, vol. 2. In: Andrés, J., Bertran, J. (eds.) Publicacions de la Universitat de Jaume I, Castelló de la Plana (2000)
77. Perdew, J.P., Zunger, A.: Phys. Rev. B **23**, 5048 (1981)
78. Tomasi, J., Persico, M.: Chem. Rev. **94**, 2027 (1994)
79. Miertus, S., Scrocco, E., Tomasi, J.: J. Chem. Phys. **55**, 117 (1981)
80. Marenich, A.V., Cramer, C.J., Truhlar, D.G.: J. Phys. Chem. B **113**, 6378 (2009)
81. Foresman, J.B., Keith, T.A., Wiberg, K.B., Snoonian, J., Frisch, M.J.: J. Phys. Chem. **100**, 16098 (1996)

Chapter 3
Objectives

All successful people men and women are big dreamers. They imagine what their future could be, ideal in every respect, and then they work every day toward their distant vision, that goal or purpose.

Brian Tracy

The main objective of this thesis is to apply computational methods to the study of Pd-catalyzed cross-coupling reactions with the aim of determining and/or better understanding their reaction mechanisms. In particular, three different Pd-catalyzed cross-coupling reactions have been investigated. A brief summary of the reasons that prompted us to study these reactions and the particular objectives established for each these studies are summarized next.

3.1 The Negishi Reaction

Despite the numerous applications of Pd-catalyzed Negishi reactions in many useful synthetic processes, such as the synthesis of fine chemicals, pharmaceuticals or natural products, the understanding of the details of its reaction mechanism is rather scarce. Thus, we decided to investigate computationally the reaction mechanism for the transmetalation process between the complex *trans*-$[Pd(Me)(Cl)(PMePh_2)_2]$ and the organozinc reagents ZnMeCl and $ZnMe_2$. This study was carried out in close collaboration with the experimental group of Prof. Pablo Espinet and Prof. Juan Casares. The main objectives of this study are:

- Determining and evaluating the transmetalation mechanisms that give rise to *trans*- and *cis*-$[Pd(Me)_2(PMePh_2)_2]$ complexes starting from *trans*-$[Pd(Me)(Cl)(PMePh_2)_2]$.
- Comparing the reaction mechanisms for the transmetalation processes with the $ZnMe_2$ and ZnMeCl reagents.

M. G. Melchor, *A Theoretical Study of Pd-Catalyzed C–C Cross-Coupling Reactions*, Springer Theses, DOI: 10.1007/978-3-319-01490-6_3,

- Investigating the effect of the additional ligands (e.g. phosphine) in the transmetalation reaction mechanism with $ZnMe_2$.
- Providing a plausible explanation for the experimental observation of the cationic complex $[Pd(Me)(PMePh_2)_3]^+$ in the transmetalation reaction with $ZnMe_2$.
- Comparing our theoretical results with the experimental results provided by the experimental group.

3.2 The Cu-Free Sonogashira Reaction

Over the last years, two reaction mechanisms proposed for the copper-free Sonogashira reaction have been somewhat discussed in the literature. Recently, the experimental group of Mårtensson demonstrated that one of them can be discarded, and further proposed two alternatives for the other mechanism on the basis of the electronic nature of the alkyne's substituents. Hence, in order to shed light on the reaction mechanism for this process, we decided to carry out a theoretical study in close contact with the experimental group of Prof. Carmen Nájera with the following main objectives:

- Evaluating all the mechanistic pathways proposed in the literature for the copper-free Sonogashira reaction.
- Exploring the possibility of new alternatives for the reaction mechanisms.
- Analyzing the effect on the reaction rate of the electronic nature of different 4-substituted phenylacetylenes (R = H, CF_3, OMe, NMe_2).
- Comparing our theoretical results with the experiments and putting them into context with the ones reported by the group of Mårtensson.

3.3 An Asymmetric Suzuki-Miyaura Reaction

Recently, the experimental research group of Prof. Rosario Fernández and Prof. José M. Lassaletta reported an asymmetric version of the Suzuki-Miyaura coupling. In particular, the use of (*S,S*)-2,5-diphenyl-pyrrolidine-derived glyoxal bis-hydrazone ligand in conjunction with Cs_2CO_3 as base and toluene as solvent, allowed the asymmetric Suzuki-Miyaura coupling of a broad variety of substrates in high yields and enantioselectivities. In order to get a deeper understanding of this reaction, we decided to carry out a theoretical study in collaboration with the group of professors Fernández and Lassaletta with the following main objectives:

- Evaluating the complete reaction mechanism for the Pd-catalyzed process using a bis-hydrazone as a ligand.
- Identifying the reaction step where the stereochemistry of the process is defined.
- Rationalizing the origin of the enantioselectivity observed for this catalyst.

Chapter 4
The Negishi Reaction Mechanism

Every science begins as philosophy and ends as art
Will Durant, The Story of Philosophy, 1926

4.1 Introduction

The Negishi reaction is a C–C cross-coupling reaction that involves the coupling between an organic halide (or triflate) R^1–X and an organozinc compound (Scheme 4.1) [1, 2]. This reaction was first reported by Prof. Negishi in 1977 showing that the reaction of aryl- and benzylzinc derivatives with aryl halides in the presence of a catalytic amount of a Ni or Pd catalyst provided a general and highly chemo- and regioselective route to unsymmetrical biaryls and diarylmethanes [3]. Thus, as most of the C–C cross-coupling reactions, the Negishi reaction is commonly catalyzed by Ni or Pd complexes, though the latter has been much more developed [4].

Most of the advantages and disadvantages of the Negishi reaction compared to other C–C cross-coupling reactions lie on the intrinsic properties of organozinc compounds. These nucleophiles, unlike other organometallic reagents such as organolithium and organomagnesium compounds, have a carbon–metal bond with a high covalent character, which makes them less reactive and, consequently, more tolerant of functional groups. Moreover, organozinc reagents have the ability of providing fast transmetalation reactions with transition metal salts, specially with those of Pd [5, 6]. Therefore, despite of their moderate reactivity toward most of the organic elec-

The results presented in this chapter have been published in the following two articles:
Article I: Fuentes, B.; **García-Melchor, M.**; Lledós, A.; Maseras, F.; Casares, J. A.; Ujaque, G.; Espinet, P. *Chem. Eur. J.* **2010**, *16*, 8596–8599.
Article II: **García-Melchor, M.**; Fuentes, B.; Lledós, A.; Casares, J. A.; Ujaque, G.; Espinet, P. *J. Am. Chem. Soc.* **2011**, *133*, 13519–13526.

M. G. Melchor, *A Theoretical Study of Pd-Catalyzed C–C Cross-Coupling Reactions*, Springer Theses, DOI: 10.1007/978-3-319-01490-6_4,

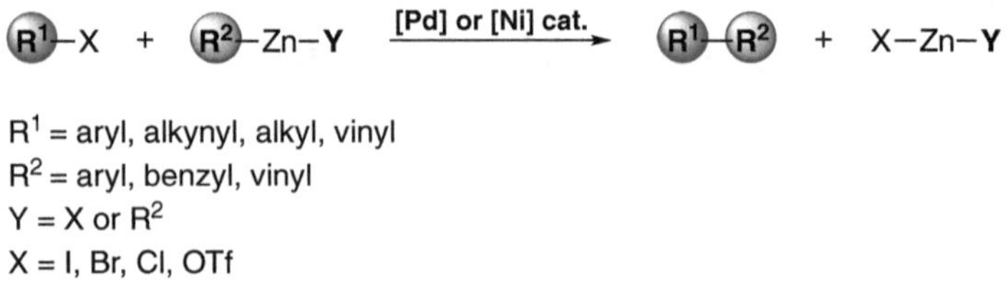

Scheme 4.1 General scheme for Negishi cross-coupling reaction

trophiles, organozinc compounds are one of the most reactive nucleophilic agents employed in Pd-catalyzed cross-coupling reactions.

Other of the main advantages of the Negishi reaction, and related to the above mentioned wide tolerance of functional groups at the organozinc reagents, is that it can be applied to every possible combination of carbon types (i.e. sp, sp^2, or sp^3). These two specific features, in particular, are responsible for the numerous applications of this reaction in many useful synthetic processes, such as the synthesis of fine chemicals, pharmaceuticals or natural products [7–11]. Furthermore, the Negishi reaction has the peculiarity that it can be carried out using two types of organometallic reagents, ZnR_2 or ZnRX. The choice of one or the other usually depends on the more readily accessibility of the two nucleophiles, though, as we will see in the next sections, the different nucleophilic character of these nucleophiles can also be considered in order to influence the reaction outcome.

As in all C–C cross-coupling reactions, the Negishi reaction mechanism consists in three steps (Fig. 4.1): oxidative addition, transmetalation, and reductive elimination. The former and the latter are common to all the other cross-coupling reactions, whereas the transmetalation step is particular of this reaction. Unfortunately, this transmetalation has been less studied compared to the ones in the Stille [12–17] or Suzuki reactions, [18–21] in spite of the fact that the transmetalation between organozinc and palladium complexes is also involved in other relevant processes, such as the hydroalkylation of styrenes, [22] the asymmetric allylation of aryl aldehydes, [23] the coupling propargylic benzoates and aldehydes, [24] or the double-transmetalation oxidative cross-coupling reaction [25, 26].

The first experimental observations on the transmetalation step in the Negishi coupling were recently reported by Espinet et al. [27] for the transmetalation reaction between *trans*-$[Pd(Rf)(Cl)(PPh_3)_2]$ (Rf = 3,5-dichloro-2,4,6-trifluorophenyl) and the organozinc reagents $ZnMe_2$ and ZnMeCl, in *THF*. Interestingly, this study revealed that each organozinc reagent affords a different isomer (*trans* or *cis*, respectively) of the coupling intermediate $[Pd(Rf)(Me)(PPh_3)_2]$, and warned on the existence of secondary undesired transmetalations[1] (i.e. methyl by aryl exchanges, Fig. 4.2) that could eventually lead to homocoupling products. This last phenomenon

[1] By "undesired transmetalations" we mean transmetalations other than the one desired to produce the heterocoupling product.

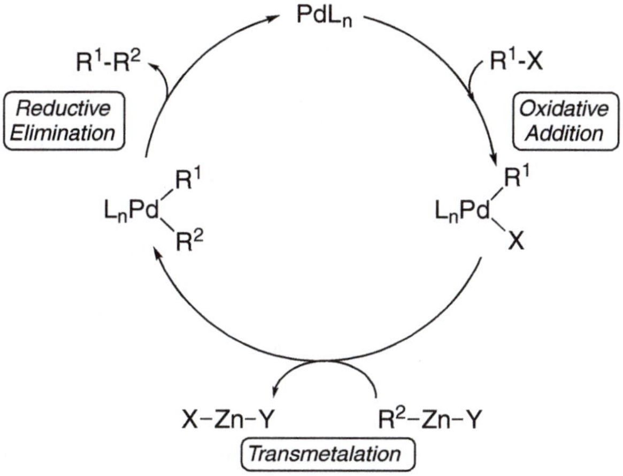

Fig. 4.1 General reaction mechanism for the Pd-catalyzed Negishi reaction

(1) *trans*-[PdMeRf(PPh$_3$)$_2$] $\xrightarrow{+\ \mathrm{ZnMe_2}}$ *cis*-[PdMe$_2$(PPh$_3$)$_2$] + ZnMeRf

(2) *trans*-[PdRfCl(PPh$_3$)$_2$] $\xrightarrow{+\ \mathrm{ZnMeCl}}$ *trans*-[PdMeCl(PPh$_3$)$_2$] + ZnRfCl

Fig. 4.2 Undesired transmetalation reactions in the Pd-catalyzed Negishi reaction between *trans*-[Pd(Rf)(Cl)(PPh$_3$)$_2$] and ZnMe$_2$ or ZnMeCl [27]

had been previously observed by Elsevier et al. [28] and, after the work of Espinet et al. [27] was also noted by Lei et al. on related reactions with aryl derivatives [29]. In particular, in this last study, the competition between a second transmetalation reaction between [Pd(Ar^1)(Ar^2)(dppf)] (dppf = 1,1'-bis(diphenylphosphino)ferrocene) and Ar^{2}ZnCl yielding [Pd(Ar^2)(Ar^2)(dppf)] and Ar^{1}ZnCl, and the reductive elimination reaction providing the coupling product Ar^1–Ar^2, was identified as responsible for the experimental observation of a mixture of heterocoupling product (Ar^1–Ar^2), homocoupling product (Ar^2–Ar^2) and dehalogenated product (Ar^1–H) (the last two in equal amounts). The theoretical calculations on a model system reported therein shed light on this issue revealing that an ortho substituent in Ar^{1}I favors the second transmetalation reaction, while an ortho substituent in Ar^{2}ZnCl significantly disfavors this second transmetalation.

It is worth mentioning that, besides the studies of Espinet et al. [27] and Lei et al. [29] above mentioned, the number of mechanistic studies reported to date concerning the Negishi reaction are more than scarce.[2] In fact, to our knowledge, there are only the ones from Organ et al. [30] and Mézailles et al. [31] The former, published the same year than Lei et al.'s paper, [29] involved the theoretical investigation of the

[2] Actually, when we started our studies on the Negishi reaction mechanism, no related studies had been reported yet.

full catalytic cycle for the Negishi Csp^3–Csp^3 coupling between EtBr and EtZnBr catalyzed by mono-ligated *NHC*–Pd complexes. Therein, the authors concluded that the use of *NHC* ligands introduce important differences into the traditionally accepted mechanism. For example, the transmetalation and not the oxidative addition, which is considered to be slow in comparison with that for aryl halides, is the rate-limiting step within the whole catalytic cycle. Moreover, the resulting inorganic salt $ZnBr_2$ from the transmetalation reaction not only does not dissociate prior to reductive elimination, but also helps to release the coupling product.

On the other hand, in the work of Mézailles et al. [31] the Negishi reaction between bromoarenes and arylzinc derivatives catalyzed by an air-stable Pd(II) complex bearing a bulky electron-withdrawing phosphine ligand was investigated both experimentally and theoretically. The *DFT* calculations on this catalytic system showed that the three elementary steps (i.e. oxidative addition, transmetalation and reductive elimination) possessed low to very low energy barriers, which agreed with the high reaction rates observed experimentally. Thus, the authors brought up the idea of replacing the bulky electron-rich phosphines, extensively used at present in cross-coupling processes, by electron-poor phosphines.

4.2 The Transmetalation Step in the Negishi Coupling of *Trans*-[PdMeCl($PMePh_2$)$_2$] with ZnMeCl

As a result of our interest in the study of cross-coupling reactions and prompted by the undesired (and unexpected) transmetalations reported by Espinet et al. [27] (Fig. 4.2), we decided to investigate computationally the reaction mechanism for the transmetalation step in the Negishi reaction. In particular, we selected the reaction of *trans*-[Pd(Me)(Cl)($PMePh_2$)$_2$] (**1**) with ZnMeCl (Scheme 4.2) because this Pd complex is analogous to the one that was supposed to be generated by a Rf-for-Me substitution in the reaction reported by Espinet et al. [27] between the starting complex *trans*-[Pd(Rf)(Cl)(PPh_3)$_2$] and ZnMeCl (Eq. 2, Fig. 4.2) [27]. Hence, the purpose of our study was to cast light on how complex **1** undergoes transmetalation with ZnMeCl, and which is the stereoselectivity of the final dimethylated product obtained in this process (Scheme 4.2).

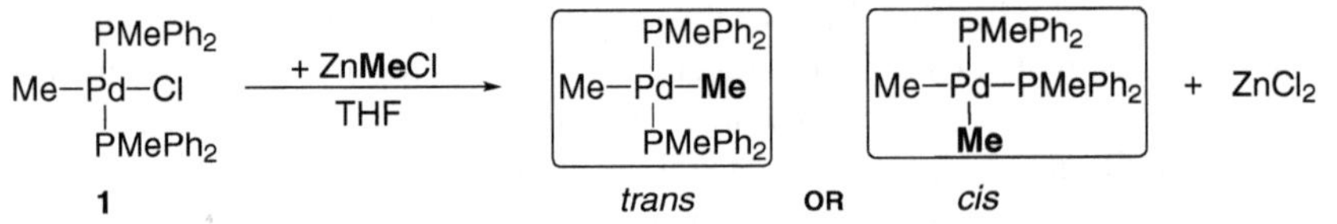

Scheme 4.2 Transmetalation reaction between *trans*-[Pd(Me)(Cl)($PMePh_2$)$_2$] (**1**) and ZnMeCl

This mechanistic study was carried out in collaboration with the experimental group of Prof. Pablo Espinet and Prof. Juan Casares at the University of Valladolid. Importantly, this study represented the first experimental determination of reaction and activation Gibbs energies for a Negishi transmetalation reaction. Furthermore, these results were contrasted with *DFT* calculations that provided a detailed picture of the reaction pathway and whose calculated energy barriers resulted to be in good agreement with the experimental parameters. In the next subsections, a summary of these experimental results followed by the theoretical analysis of this work will be presented.

4.2.1 Experimental Data

As mentioned, all the experiments in this study were performed in the research group of Prof. Espinet and Prof. Casares at the University of Valladolid. The reactions between **1** and ZnMeCl (Scheme 4.2) were carried out (with one exception) in 1:20 ratio, simulating catalytic conditions with 5 % of Pd in *THF*, and at different temperatures.

Interestingly, experiments revealed significant differences between the transmetalation reactions carried out at different temperatures. Thus, in the transmetalation reaction at room temperature the only product that could be observed (in equilibrium with the starting complex **1**) was the complex *cis*-$[Pd(Me)_2(PMePh_2)_2]$ (**2**), which slowly underwent reductive elimination producing ethane. In contrast, when the same reaction was monitored by ^{31}P NMR spectroscopy at 223 K (Fig. 4.3a), the coupling rate to yield ethane became negligible and the formation of *trans*-$[Pd(Me)_2(PMePh_2)_2]$ (**3**) as well as *cis*-$[Pd(Me)_2(PMePh_2)_2]$ (**2**) was observed. Furthermore, at this temperature, the *trans* product **3** seemed to be formed first and then fade out in favor of **2**. Hence, in order to get deeper insight into this exciting finding, the same reaction but now at 203 K and in 1:1 ratio to get a slower rate of transformation, was also carried out (Fig. 4.3b). The transmetalation reaction monitored by ^{31}P NMR in these conditions confirmed that **3** is formed noticeably faster than **2**, which allowed us to conclude that the only observation of the *cis* product **2** at room temperature is deceptive for the stereoselectivity of the transmetalation.

The behavior of **3** found in these experiments resembles to the one of a kinetic product of considerably lower stability than the thermodynamic product (**2**). That is, it disappears from observation as the reaction evolves and gets closer to the equilibrium concentrations, where the concentration of **3** is very small. Indeed, this can be observed from the concentration versus time data obtained by monitoring the reaction at 223 K (Fig. 4.3a): the concentration of **2** increases continuously, whereas a small accumulation of **3** is initially produced followed by a decrease in its concentration, so that after 300 min it has practically disappeared.[3]

[3] After about 10 h at this temperature, the system has reached equilibrium between the starting material **1** and the final thermodynamic product **2** ([**1**] = 5.8×10^{-3} M, [**2**] = 4.4×10^{-3} M, and

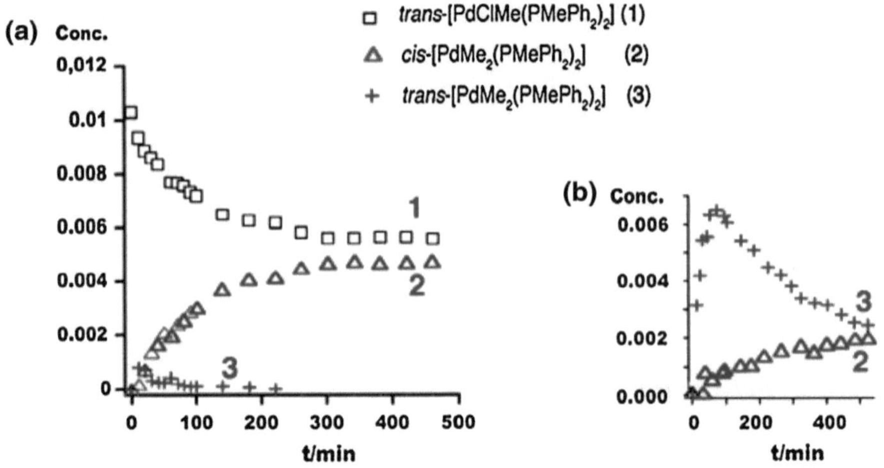

Fig. 4.3 Concentration versus time data obtained by ^{31}P NMR spectroscopy for the reaction of **1** with ZnMeCl in different conditions: **a** ratio 1:20, in *THF* at T = 223 K. Starting conditions: $[\mathbf{1}]_0$ = 0.01 M; $[\text{ZnMeCl}]_0$ = 0.20 M; **b** ratio 1:1, in *THF* at T = 203 K (in this case **1** is not plotted because its abundance is above the values represented in the ordinate axis). Starting conditions: $[\mathbf{1}]_0$ = 0.056 M; $[\text{ZnMeCl}]_0$ = 0.056 M

Another surprising finding was the fact that despite the transient existence of **3** in the transmetalation reactions, this species prepared by an alternative method was found to be fairly stable. In particular, it took 10 h in *THF* at 273 K for about half of it to isomerize to **2**, whereas at 223 K the **3** to **2** isomerization rate was negligible. On the other hand, the addition of $ZnCl_2$ to a solution of **3** in *THF* at 223 K, produced the instantaneous and complete transformation of **3** to **1**. In the same conditions (i.e. addition of $ZnCl_2$), compound **2** was also transformed into **1** until equilibrium was reached. In other words, the transmetalations between **1** and ZnMeCl to give **2** or **3** are quickly reversible. Hence, these experiments demonstrated that the fast **3** to **2** isomerization observed in Fig. 4.3 is not a direct isomerization, which is slow, but a retrotransmetalation of **3** to **1** followed by transmetalation to **2**.

Finally, the concentration versus time data for the reaction monitored at 223 K (Fig. 4.3a) were fitted to the kinetic model shown in Fig. 4.4, thereby allowing the calculation of the transmetalation and retrotransmetalation rates at this temperature.[4] Then, from these rates, the relative Gibbs energies at that temperature ($\Delta G_{223\,K}$) for **1**, **2**, **3** and for the transition states that connect these species ($\mathbf{TS_{1-2}}$ and $\mathbf{TS_{1-3}}$) were obtained (Fig. 4.4). The calculated Gibbs energies indicate that the transmetalation from **1** to **3** requires 1 kcal · mol^{-1} less than the transmetalation from **1** to **2**. This, in terms of reaction rates, means that the transmetalation from **1** to **3**

K_{eq} = 2.0×10^{-2}). The concentration of **3** is not given because it is below the limit of NMR observation.

[4] This fitting was also performed by the group of Prof. Espinet and Prof. Casares using the multivariable adjust program *Gepasi* [32].

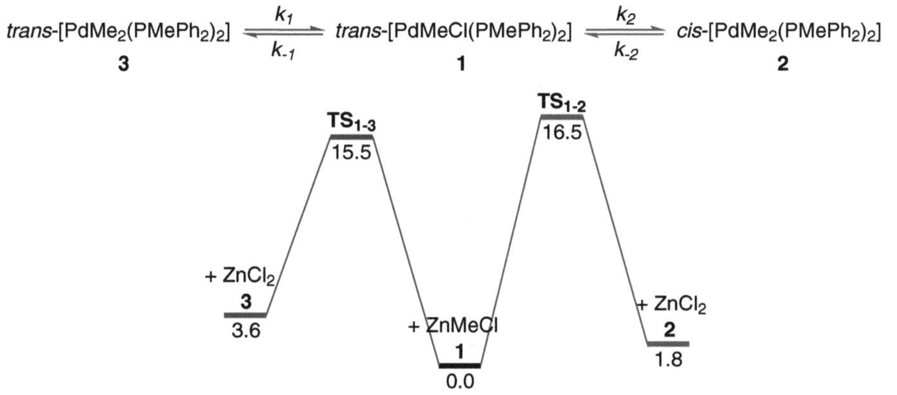

Fig. 4.4 Kinetic model (*top*) and experimental $\Delta G_{223\,K}$ profiles (*bottom*) (kcal · mol^{-1}) for the reaction of **1** with ZnMeCl. k_1 and k_2 are the transmetalation rate constants, whereas k_{-1} and k_{-2} are the retrotransmetalation rate constants

(at 223 K) is about 10 times faster than that to **2**. Furthermore, the values indicate that the retrotransmetalations from **3** or **2** to **1** are still faster than their respective transmetalations.

It is noteworthy to mention that the fact that the experimental results have been summarized first does not necessarily mean that all the events in this study followed this particular chronological order. In fact, experiments and theoretical calculations were carried out hand in hand. Thereby, the proposal of additional experiments and/or calculations during the course of this study was, in many cases, instigated by theoretical and experimental evidences, respectively. Consequently, many of the results that have been described in this section (or that will be described in further sections) are fruit of the rationalization of a series of experimental and theoretical findings.

In the following sections, the computational details of the theoretical calculations performed in this study, followed by the results derived from those, will be described.

4.2.2 Computational Details

All the theoretical calculations in this study were carried out at the *DFT* level by means of the dispersion-corrected *M06* functional [33] and using the *Gaussian09* program [34]. We selected this functional in purpose because it has been shown to deliver good results for both organometallic systems and non-covalent interactions [33]. The mechanistic study was performed on the very Pd catalyst used for experiments, thus without simplifying the phosphine ligands bound to the metal center. Moreover, for the ZnMeCl reactant two additional explicit solvent molecules (i.e. *THF*) were included to fulfill the coordination sphere of Zn. The choice of this methodology allowed us to consider tetrahedral Zn species coordinated with *THF* along all the studied reaction pathways, rather than the unrealistic linear dicoordinated Zn species often used and that lead to calculated species of unlikely existence in *THF* solution [35].

For geometry optimizations, the standard 6–31G(d) basis set was used to describe the C, Cl, O, P, and H atoms, and the *LANL2DZ* effective core potential [36] for the Pd and Zn atoms. We denoted the combination of these two basis sets as *BSI*. Geometries of all the reactants, intermediates, and transition states were fully optimized without symmetry constraints. The nature of these stationary points was characterized by a vibrational analysis performed within the harmonic approximation. In particular, transition states were identified by the presence of one imaginary frequency and minima by a full set of real frequencies. Each of the computed reaction pathways was further checked by connecting transition states to the corresponding reactants and products following the displacement associated to the imaginary frequencies.

Potential energies (both in gas phase and solution) were also obtained by performing single point calculations at the above optimized geometries using the all electron basis set *DGDZVP* (double-ζ valence plus polarization) [37] for Pd, and the *TZVP* basis set (triple-ζ valence plus polarization) [38] for Zn and the main group elements. We denoted the combination of these two basis sets as *BSII*. Relative Gibbs energies were obtained by adding to the potential energies calculated with *BSII* the zero point energy (*ZPE*) and the vibrational, rotational, and translational entropic contributions of all the species calculated with *BSI* and at the temperature of 223 K.

Solvent effects (i.e. *THF*, $\varepsilon = 7.4257$) were introduced by a discrete-continuum model: two *THF* molecules were explicitly included in the calculations as potential ligands (see above), and the effect of the bulk solvent was considered with a continuum method, the *PCM* approach [39], by means of single point calculations at all optimized gas phase geometries. In this method, the radii of the spheres employed to create the cavity for the solute were defined with the *UFF* model,[5]. which is the default in *Gaussian09*.

Unless otherwise specified, all the energies shown in this study correspond to Gibbs energies in *THF* (ΔG_{THF}) at 223 K, obtained by employing the following scheme:

$$\Delta G_{THF} = \Delta E_{THF} + (\Delta G_{gas} - \Delta E_{gas})$$

4.2.3 The Concerted Mechanism to the Trans Intermediate

As commented, experiments revealed with the reaction of **1** with ZnMeCl at 203 K that the *trans* product (**3**) is formed at a higher rate than the *cis* one (**2**), thus demonstrating that the unique observation of **2** when the reaction is carried out at room temperature is deceptive. Hence, with the aim of providing a better understanding of this complex transmetalation process, we first focused our attention on

[5] For further details on this model, see the last section of Chap. 2

computing a reaction pathway for the transmetalation of **1** leading to *trans* product **3**. In particular, we computed a reaction mechanism similar to the typical exchange-type mechanism described in Chap. 1 and that entails the concerted ligand exchange between two metal centers. The Gibbs energy profile obtained for this reaction pathway is depicted in Fig. 4.5.

This concerted mechanism starts with the dissociation of a *THF* molecule from the organozinc reagent $ZnMeCl(THF)_2$ and the concomitant bridging chloride coordination to Zn. The resulting intermediate from this first step is **CT-1**, with an energy of 3.1 kcal · mol^{-1} relative to reactants. In this first intermediate, the Pd complex adopts a square planar geometry, whereas the Zn reagent adopts a Y-type geometry placed at 3.397 Å from Pd and 2.562 Å from the Cl atom (Fig. 4.6). Then, from this species, the concerted transmetalation reaction occurs with a relative energy barrier of only 7.4 kcal · mol^{-1}. This step takes place through the four-membered ring transition state **CT-TS1** and involves the simultaneous methyl and chloride exchange between the Zn and Pd centers. Thus, in **CT-TS1** the two Pd–Cl (2.883 Å) and Zn–Me (2.071 Å) bonds are being broken at the same time that the two new Pd–Me (2.794 Å) and Zn–Cl (2.391 Å) bonds are being formed.

Interestingly, the intermediate that results from the concerted transmetalation (**CT-2**) features a relatively short Pd–Zn distance (2.710 Å), which suggests the existence of a metal-metal interaction between the electron rich Pd center and the fairly positive

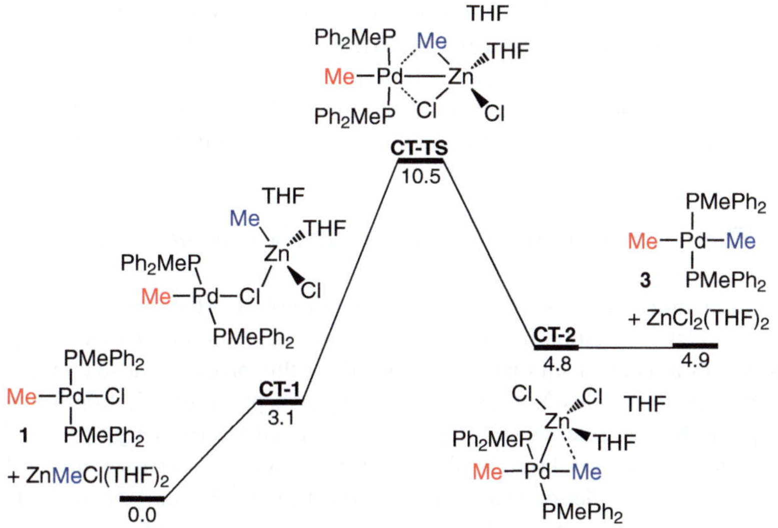

Fig. 4.5 Gibbs energy profile in *THF* (ΔG_{THF}, kcal · mol^{-1}) at 223 K for the transmetalation with ZnMeCl leading to the *trans* product (**3**) via a concerted mechanism

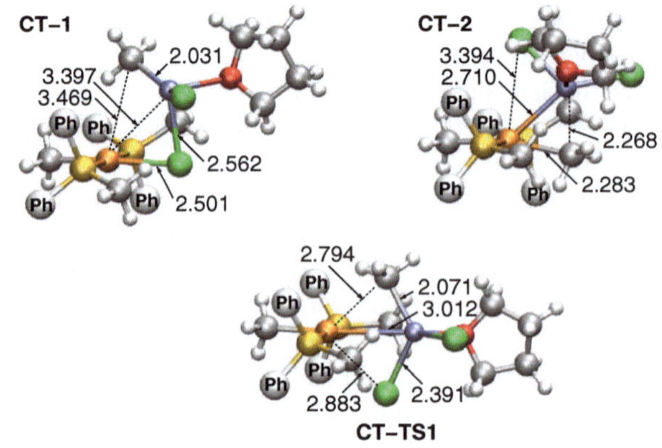

Fig. 4.6 Optimized structures for the transition state and all the intermediates involved in the transmetalation with ZnMeCl leading to the *trans* product (**3**) via a concerted mechanism. Phenyl rings of the phosphine ligands have been simplified, and the second *THF* molecule of the organozinc reactant is not shown for clarity. Distances are shown in Å

Zn center.[6] Once **CT-2** is formed, this last intermediate affords the final product **3** by coordination of a *THF* molecule to the Zn reagent and subsequent Pd–Zn dissociation.

The optimized geometries for the transition state and the intermediates involved in this concerted mechanism are shown in Fig. 4.6.

Overall, the computed concerted mechanism for the transmetalation reaction of **1** to the *trans* product **3** (Fig. 4.5) shows that this process is endergonic by 4.9 $kcal \cdot mol^{-1}$. Therefore, the reaction equilibrium is displaced toward the starting materials (**1** + $ZnMeCl(THF)_2$). On the other hand, the energy difference between the products (**3** + $ZnCl_2(THF)_2$) and the starting materials is responsible for the lower energy barrier for the reverse reaction (i.e. retrotransmetalation from **3** to **1**) compared to the forward reaction (i.e. transmetalation from **1** to **3**). This difference in the energy barriers reflects in a faster retrotransmetalation compared to transmetalation.

4.2.4 The Concerted Mechanism to the cis *Intermediate*

With the reaction mechanism for the transmetalation of **1** to the *trans* product **3** established, we next focused on investigating the transmetalation of **1** to the *cis* product **2**. The reaction mechanism that we computed for this process is also concerted but, unlike the one that yields **3**, it involves a Me by phosphine substitution followed by a phosphine by chloride substitution. The calculated Gibbs energy profile for the transmetalation reaction through this concerted mechanism is shown in Fig. 4.7.

As in the transmetalation leading to the *trans* product **3** (Fig. 4.5), the first step in this mechanism corresponds to the dissociation of a *THF* molecule from the organozinc reagent $ZnMeCl(THF)_2$ and the concomitant bridging chloride coordination to Zn (Fig. 4.7). The intermediate that results from this first step is **CC-1** (analogous

[6] Such Pd–Zn interactions were also noticed by Álvarez et al. [40] in the formation of nucleophilic allenylzincs from allenylpalladium complexes, and by Organ et al. [30] in the alkyl-alkyl Negishi coupling catalyzed by (*NHC*)-Pd complexes.

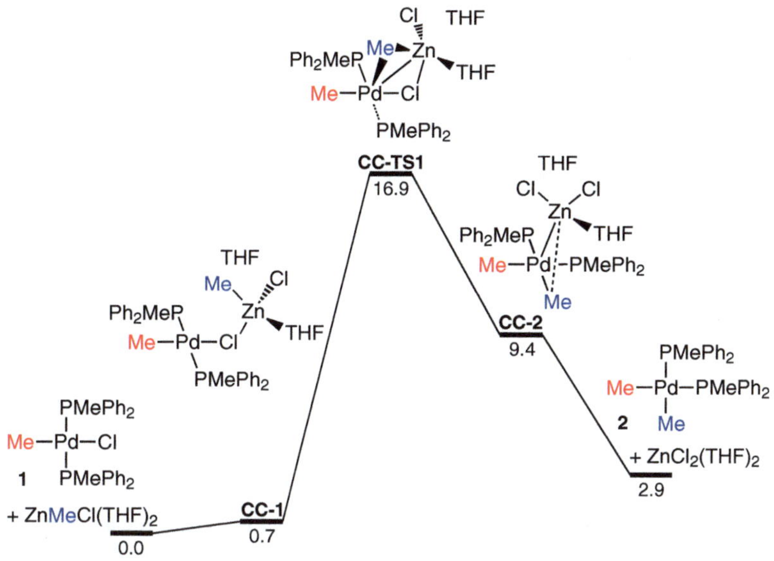

Fig. 4.7 Gibbs energy profile in *THF* (ΔG_{THF}, kcal · mol^{-1}) at 223 K for the transmetalation with ZnMeCl leading to the *cis* product (**2**) via a concerted mechanism

to **CT-1**),[7] which undergoes Me by phosphine substitution followed by phosphine by chloride substitution. Interestingly, this double substitution was found to take place in a unique transition state (**CC-TS1**) with an energy barrier of 16.9 kcal · mol^{-1}, though, as we will see later, it can also occur in two subsequent steps (i.e. transmetalation of **1** with $ZnMe_2$).

Similarly to **CT-TS1** in Fig. 4.6, **CC-TS1** features a four-membered ring formed between the Pd, Cl, Me and Zn atoms (Fig. 4.8). Moreover, the Pd–Zn distance in this transition state is rather short (2.592 Å), which seems to indicate that indeed there exists an interaction between these two metal centers. This transition state evolves to intermediate **CC-2**, in which the Pd–Zn interaction is maintained with a distance of 2.586 Å. Finally, **CC-2** affords the *cis* product **2** by coordination of a *THF* molecule to the Zn reagent and subsequent Pd–Zn dissociation. The optimized geometries for all the species that take part in this concerted mechanism are shown in Fig. 4.8.

According to the overall Gibbs energy profile presented in Fig. 4.7, the transmetalation of **1** to the *cis* product **2** is endergonic by 2.9 kcal · mol^{-1}. Therefore, just like in the transmetalation of **1** to **3**, the reaction to **2** is displaced toward the starting reagents (**1** + $ZnMeCl(THF)_2$). Furthermore, this energy difference between reactants and products also causes the rate of retrotransmetalation to be higher than that for transmetalation.

[7] The difference in energy between **CT-1** and **CC-1** stems from the different conformation of the Zn reactant in these complexes.

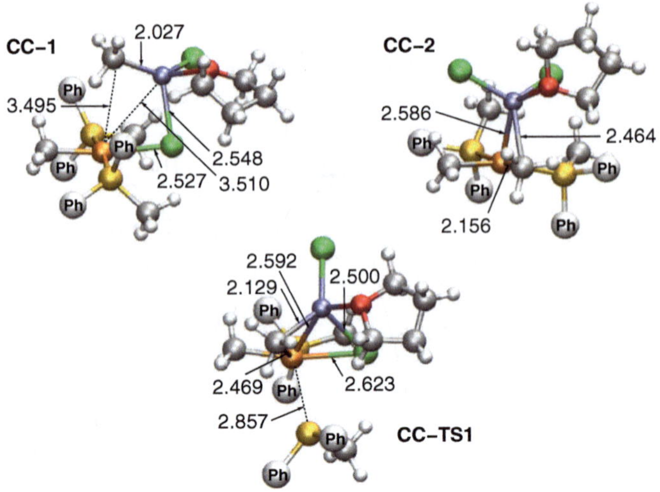

Fig. 4.8 Optimized structures for the transition state and all the intermediates involved in the transmetalation with ZnMeCl leading to the *cis* product (**2**) via a concerted mechanism. Phenyl rings of the phosphine ligands have been simplified, and the second *THF* molecule of the organozinc reactant is not shown for clarity. Distances are shown in Å

4.2.5 Comparing Theoretical and Experimental Results

So far, the experimental and theoretical results obtained for the transmetalation reactions of **1** to **2** or **3** have been presented, but no comparison between them has been made yet. Hence, with this aim, the Gibbs energies obtained in the theoretical and experimental studies for these two transmetalations at 223 K have been summarized in the simplified reaction profiles depicted in Fig. 4.9. According to these results, the theoretical calculations reproduce qualitatively all the experimental observations. In particular, from the thermodynamic point of view, both computed transmetalation reactions are endergonic, in agreement with experiments. Furthermore, the order of stability of the reagents and products in equilibrium predicted by calculations is exactly the same that the one observed in the experiments: **1** > **2** (*cis*) > **3** (*trans*). Therefore, calculations also conclude that **2** is the thermodynamic product.

As far as the kinetics is concerned, the theoretical results show that the reverse reactions (i.e. retrotransmetalations from **2** or **3** to **1**) are faster, which means that both transmetalation reactions are quickly reversible. This is consistent with the experimental observation that the addition of $ZnCl_2$ to a solution of **2** or **3** leads to the transformation of these species to **1**, and accordingly, with the fact that the isomerization of **3** to **2** occurs via retrotransmetalation from **3** to **1** and subsequent transmetalation to **2**. On the other hand, the computed energy barriers for both reaction pathways are low, which accounts for the fast transmetalation reactions observed in the experiments. More specifically, the global energy barrier for the transmetalation reaction to the *trans* product **3** ($10.5\,\text{kcal}\cdot\text{mol}^{-1}$) is lower than the one to the

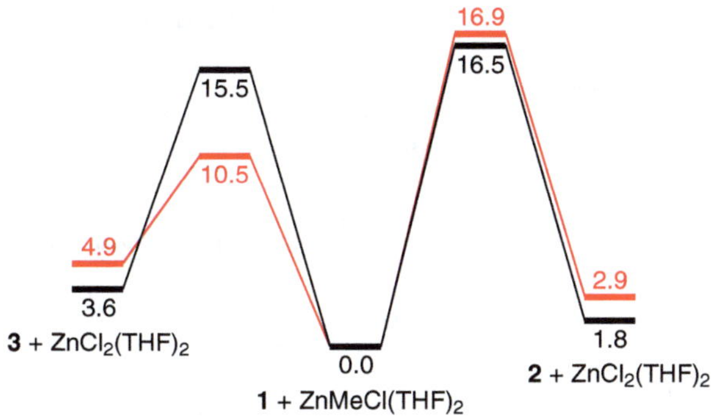

Fig. 4.9 Comparison of the Gibbs energy values at 223 K ($\Delta G_{223\,K}$, kcal · mol^{-1}) for the transmetalation reaction of **1** with ZnMeCl. Experimental values are shown in *black*, whereas calculated values are shown in *red*

cis product **2** (16.9 kcal · mol^{-1}), which indicates that **3** is the kinetic product in agreement with experiments.

From the quantitative perspective, the results collected in Fig. 4.9 show a reasonable match between the computed and experimental values for the stabilities of the reagents and products in equilibrium (with energy differences about 1 kcal · mol^{-1}). On the other hand, the computed Gibbs energy barriers show an excellent fit with the experimental value for the transmetalation of **1** to **2** (with a difference of less than 0.5 kcal · mol^{-1}), but not so good for the transmetalation to **3** (calculations underestimates the Gibbs energy barrier by 5 kcal · mol^{-1}).

4.2.6 Conclusions

The transmetalation reaction between the complex *trans*-[Pd(Me)(Cl)($PMePh_2$)$_2$] (**1**) and ZnMeCl (Scheme 4.2) was investigated combining experiments and *DFT* calculations. The theoretical and experimental results derived from this study revealed that, at variance with the unique observation of the *cis* product (**2**) when the reaction is carried out at room temperature, the *trans* product (**3**) is formed at a higher rate. Thus, the reaction of **1** with ZnMeCl follows two competitive pathways: one, kinetically preferred that yields **3** but is unproductive for the coupling; and other, about one order of magnitude slower, that produces **2** from which the coupling will eventually take place though at a much slower rate. The computation of these two pathways showed that both involve cyclic transition states with structures very reminiscent of the one proposed for the cyclic mechanism in the Stille reaction [12, 14]. However, in the case of the Negishi reaction a distinctive feature of these transition states and some intermediates was found: the existence of metal-metal interactions between the Pd and Zn centers.

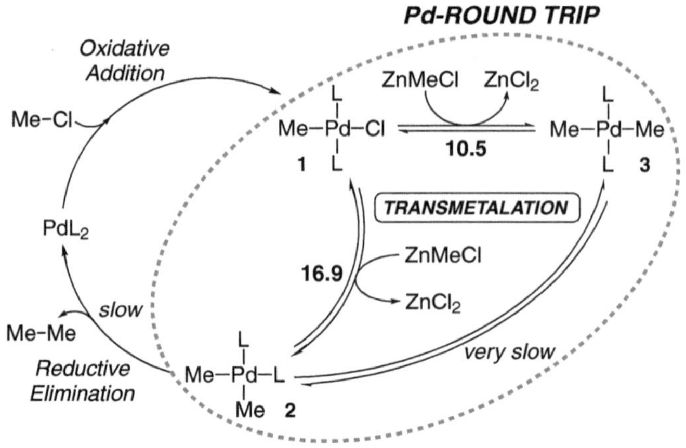

Fig. 4.10 Simplified reaction mechanism for the Pd-catalyzed Negishi cross-coupling reaction with ZnMeCl (*L* = $PMePh_2$)

Finally, this study also demonstrated that the transmetalations of **1** to **2** or **3** are quickly reversible, whereas the isomerization of **3–2** is very slow. Hence, since the transmetalation reactions and their respective reverse processes are much faster than the reductive elimination from **2**, it follows that the **2–1–3** round trip will take place many times before coupling occurs. Accordingly, in general catalyzed couplings of $[Pd(R^1)(X)(L)_2]$ and R^2ZnCl (or ZnR^2_2) there are statistically many chances for new intermediates $[Pd(R^2)(X)(L)_2]$ and R^1ZnCl (or ZnR^1R^2) to be formed through undesired retrotransmetalations from the initial intermediate $[Pd(R^1)(R^2)(L)_2]$, eventually affording the R^1–R^1 and R^2–R^2 homocoupling side products. All these results are summarized in the catalytic cycle shown in Fig. 4.10.

Overall, this study highlighted the complexity of the transmetalation step in the Negishi reaction. It is noteworthy to mention that this study represented the first experimental determination of reaction and activation Gibbs energies for a Negishi transmetalation reaction, and one of the earliest theoretical studies on this particular step. Furthermore, the deeper understanding provided by this study of the transmetalation of **1** with ZnMeCl is of particular importance since this type of complex can be envisaged in general catalyzed couplings of $[Pd(R^1)(X)(L)_2]$ and R^2ZnCl (or ZnR^2_2) as a result of undesired transmetalations.

4.3 The Transmetalation Step in the Negishi Coupling of *Trans*-$[PdMeCl(PMePh_2)_2]$ with $ZnMe_2$

Due to the fact that the Negishi reaction is peculiar in that it can be carried out with two different types of organozinc nucleophiles (i.e. ZnR_2 and ZnRX), we decided, for the sake of completeness, to study the same transmetalation reaction that we investigated with ZnMeCl, [41] but with $ZnMe_2$ (Scheme 4.3).

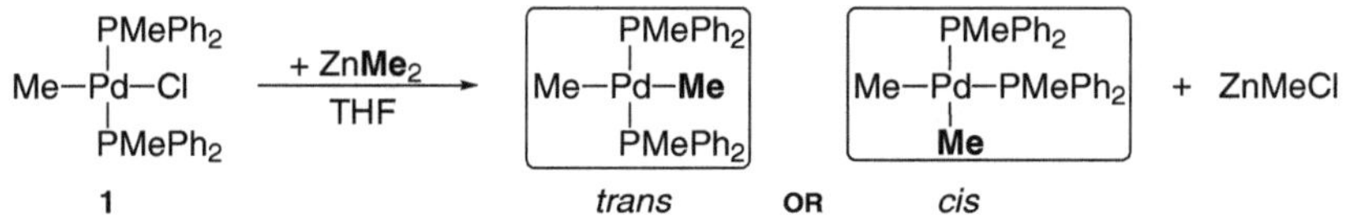

Scheme 4.3 Transmetalation reaction between *trans*-[Pd(Me)(Cl)($PMePh_2$)$_2$] (**1**) and $ZnMe_2$

This study was also realized in close collaboration with the experimental research group of Prof. Espinet and Prof. Casares. Hence, the reaction mechanism for the transmetalation of (**1**) with $ZnMe_2$ was investigated both theoretically and experimentally. As for the transmetalation with ZnMeCl, in the following sections the experimental and theoretical results derived from this study will be presented in this order.

4.3.1 Experimental Data

A priori, with the merely change of ZnMeCl by $ZnMe_2$ as transmetalating nucleophile, only minor differences with respect to the results obtained with ZnMeCl were expected. However, as we will see next, experiments and theoretical calculations on this reaction revealed several unknown and unexpected aspects of the Negishi reaction.

As for the transmetalation with ZnMeCl, the reactions between **1** and $ZnMe_2$ (Scheme 4.3) were carried out in 1:20 ratio, simulating catalytic conditions with 5 % of Pd in *THF*, and at different temperatures. In coincidence with ZnMeCl, the reaction with $ZnMe_2$ at room temperature afforded complex **2** as the only observable product, which slowly decomposed to give the Negishi coupling product ethane. In contrast, when the reaction was monitored by ^{31}P NMR at 203 K,[8] significant differences between the two organozinc reagents were observed (Fig. 4.11). For example, the rate of consumption of **1** with $ZnMe_2$ (Fig. 4.11a) resulting in a mixture of **3** and **2** (being **3** very major) seemed to be too high compared to that with ZnMeCl, even considering that some acceleration was expected owing to the higher nucleophilic character of $ZnMe_2$. Besides, unlike the reaction with ZnMeCl, where the reaction equilibrium is displaced toward the reagents (**1** + ZnMeCl, see Fig. 4.3), in the system **1** + $ZnMe_2$ the reaction is displaced toward the products (**2** + ZnMeCl, and **3** + ZnMeCl). On the other hand, the addition to the reaction of just a small amount of $PMePh_2$ markedly decreased the rate of consumption of **1** (Fig. 4.11b), thus bringing the reaction rate into the convenient range for kinetic studies with ^{31}P NMR. Moreover, interestingly enough, further increases in the concentration of $PMePh_2$ affected only slightly the rate of consumption of **1** (Fig. 4.11c–e).

[8] Temperature at which the coupling rate to give ethane was negligible.

Hence, the behavior observed in the reactions with $ZnMe_2$ pointed out to the existence, in the initial reaction conditions (Fig. 4.11a), of a minute proportion of a non-observed catalytic intermediate that would open a much faster reaction pathway; this intermediate with the addition of a small amount of phosphine would be blocked, thereby causing the reaction rate to decrease. As the most plausible candidate for such catalytic intermediate we proposed the cationic species *trans*-$[Pd(Me)(PMePh_2)_2(THF)]^+$ (**4**$^+$), which would be generated in the presence of a large amount of $ZnMe_2$ acting as Cl^- scavenger.[9] This proposal was reinforced by the observation of the related cationic complex $[Pd(Me)(PMePh_2)_3]^+$ (**5**$^+$) in the reactions with added $PMePh_2$ (Fig. 4.11b–e), and further supported by *DFT* calculations (see later).

Interestingly, the observation of **5**$^+$ in variable concentration depending on the amount of added phosphine confirmed unequivocally that, in the presence of an excess of phosphine ligand, a second cationic species, this time observable, is made available from the very beginning of the reaction. Therefore, depending on the reaction conditions, the transmetalation with $ZnMe_2$ could take place on **1**, **4**$^+$, or on **5**$^+$. In order to shed light on this issue, the kinetic data collected in Fig. 4.11 was fitted to two kinetic models[10]: (a) considering that the transmetalation takes place only from **1** (Eqs. 1 and 2, Fig. 4.12); and (b) taking also into account the competitive transmetalation occurring from the cationic complex **5**$^+$ (Eqs. 3 and 4, Fig. 4.12).

Importantly, the results derived from the kinetic model *b* provided negligible values for k_3^+, thus indicating that the competitive transmetalation pathway via **5**$^+$ can be ruled out. Besides, both kinetic models consistently provided almost identical values for k_3 and for k_4.[11]

Hence, according to the experiments, in the absence of added phosphine the observed rate is too fast for the transmetalation to occur on **1**; consequently, the reaction must be going via the non-observed cationic intermediate **4**$^+$. Similarly to the reaction with ZnMeCl, the transmetalation reaction affords the *trans* product **3** at a higher rate than the *cis* product **2**. Furthermore, the isomerization of **3** to **2** is also slow. On the other hand, in the presence of added phosphine, the cationic intermediate **4**$^+$ is not available and, accordingly, the transmetalation takes place on **1**, but at a lower rate.

As commented for the reaction with ZnMeCl, the experiments and theoretical calculations in this study were performed alongside each other. Thus, although the experimental results have been presented first, it should be said that many of the mechanistic suggestions or proposals presented in this section were fruit of the profitable feedback between experiments and theory.

---

[9] Precedents for this proposal are the role of $ZnBr_2$ facilitating halide abstraction from Ni complexes, reported by Buchwald, [42, 43] and the formation of $[Ni(\eta^3\text{-Bz})(\text{diphosphine})][ZnBr_3(THF)]$ complexes from $[Ni(\sigma\text{-Bz})Br(\text{diphosphine})]$ and $ZnBr_2$ in *THF*, reported by Anderson and Vicic [44].

[10] For further details see Tables 1 and 2 in Ref. [45].

[11] The two rate constants with average values $k_3 \approx 2.8\times10^{-2}$ min^{-1}, and $k_4 \approx 6.2\times10^{-3}$, differ in less than 1 order of magnitude.

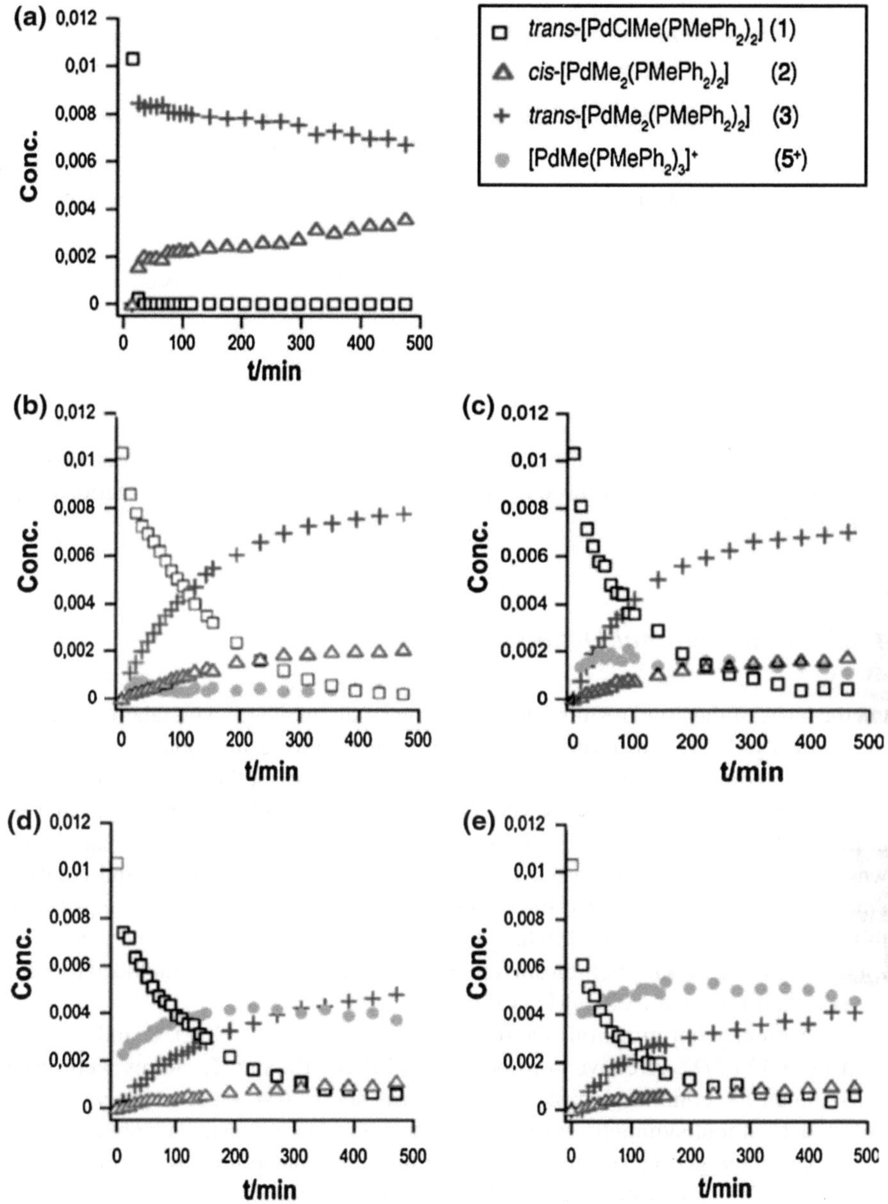

Fig. 4.11 Concentration versus time data obtained by ^{31}P NMR spectroscopy for the reaction of **1** with $ZnMe_2$ in *THF* at T = 203 K. Starting conditions: $[\mathbf{1}]_0$ = 0.01 M; $[ZnMe_2]_0$ = 0.21 M. Added phosphine: **a** $[PMePh_2]$ = 0 M, **b** $[PMePh_2] = 6.0\times10^{-4}$ M, **c** $[PMePh_2] = 2.0\times10^{-3}$ M, **d** $[PMePh_2] = 5.0\times10^{-3}$ M, **e** $[PMePh_2] = 1.0\times10^{-2}$ M

$$(1)\quad \textit{trans}\text{-}[PdMeClL_2]\ (\mathbf{1}) + ZnMe_2 \underset{k_{-3}}{\overset{k_3}{\rightleftharpoons}} \textit{trans}\text{-}[PdMe_2L_2]\ (\mathbf{3}) + ZnMeCl$$

$$(2)\quad \textit{trans}\text{-}[PdMeClL_2]\ (\mathbf{1}) + ZnMe_2 \underset{k_{-4}}{\overset{k_4}{\rightleftharpoons}} \textit{cis}\text{-}[PdMe_2L_2]\ (\mathbf{2}) + ZnMeCl$$

$$(3)\quad \textit{trans}\text{-}[PdMeClL_2]\ (\mathbf{1}) + ZnMe_2 + L \underset{k_{-5}}{\overset{k_5}{\rightleftharpoons}} [PdMeL_3]^{\oplus}[ZnMe_2Cl]^{\ominus}\ (\mathbf{5^+})$$

$$(4)\quad [PdMeL_3]^{\oplus}[ZnMe_2Cl]^{\ominus}\ (\mathbf{5^+}) \underset{k^+_{-3}}{\overset{k^+_3}{\rightleftharpoons}} \textit{trans}\text{-}[PdMe_2L_2]\ (\mathbf{3}) + ZnMeCl + L$$

Fig. 4.12 Reactions considered for the two applied kinetic models

In the following sections, the computational details of the theoretical calculations performed in this study, followed by the results derived from these calculations, will be presented.

4.3.2 Computational Details

On the basis of the good results obtained in the study on the transmetalation reaction of **1** with ZnMeCl, and in order to further compare the results obtained with both organozinc compounds, we adopted essentially the same methodology than for that study. Hence, all the calculations were performed with the *Gaussian09* program [34] at the *DFT* level and by means of the *M06* functional [33]. The mechanistic study was also carried out considering the same molecules (i.e. reactants and catalyst) employed for the experiments, thus without modeling the phosphine ligands bound to the Pd catalyst (i.e. methyldiphenylphosphine). Moreover, as for the study with ZnMeCl, two additional explicit solvent molecules (i.e. *THF*) were included to fulfill the coordination sphere of $ZnMe_2$ [35].

For geometry optimizations the 6–31G(d) basis set for C, Cl, O, P, and H, along with the *LANL2DZ* effective core potential [36] for Pd and Zn atoms were used. We denoted the combination of these two basis set as *BSI*. Geometries were fully optimized without symmetry constraints. Harmonic force constants were computed at the optimized geometries to characterize the stationary points as minima or saddle points. The latter were confirmed by having a unique imaginary frequency and correlating the corresponding reactants and products.

Potential energies (both in gas phase and in solution) were refined by performing single point calculations at the optimized geometries with *BSI* using the all electron basis set *DGDZVP* (double-ζ valence plus polarization) [37] for Pd, in conjunction with the *TZVP* (triple-ζ valence plus polarization) [38] basis set for Zn and the main group elements. We named the combination of these two basis sets as *BSII*. Relative Gibbs energies in *THF* of the species were obtained by adding the gas phase

Gibbs energy corrections of the solute calculated with *BSI* at 203 K to the energies in solution calculated with *BSII*.

Solvent effects (i.e. *THF*, $\varepsilon = 7.4257$) were introduced by a discrete-continuum model: two *THF* molecules were explicitly included in the calculations to fulfill the coordination sphere of the Zn reagent, and the effect of the bulk solvent was taken into account with a continuum method, the *SMD* [46] solvation model[12] implemented in *Gaussian09*, by means of single point calculations at all optimized gas phase geometries.

All the energies shown in this study correspond to Gibbs energies in *THF* (ΔG_{THF}) at 203 K calculated by employing the following scheme:

$$\Delta G_{THF} = \Delta E_{THF} + (\Delta G_{gas} - \Delta E_{gas})$$

4.3.3 The Concerted Mechanism to the Trans *Intermediate*

The mechanistic study on the transmetalation of **1** with ZnMeCl revealed that the *trans* product **3** is formed through a reaction pathway that involves the concerted Cl for Me exchange in a cyclic transition state (Fig. 4.5). The analogous mechanism for the reaction of **1** with $ZnMe_2$ was also computed. The calculated Gibbs energy profile for the reaction occurring via this mechanism is depicted in Fig. 4.13.

Similarly to the concerted pathway with ZnMeCl, the first step in this mechanism corresponds to the dissociation of a *THF* molecule from the starting Zn reagent and the bridging chloride coordination to Zn. This initial step is almost thermoneutral ($0.8\,\text{kcal}\cdot\text{mol}^{-1}$) and results in the formation of the intermediate **CT-1**. Subsequently, this species evolves to the transition state **CT-TS1**, where the transmetalation properly speaking occurs with a global energy barrier of $10.7\,\text{kcal}\cdot\text{mol}^{-1}$. The optimized geometry for **CT-TS1** is practically the same than the calculated for the transition state with ZnMeCl (**CT-TS1**, Fig. 4.6), with distances differing in less than about 0.1 Å (Fig. 4.14). This transition state affords the last intermediate **CT-2**, which features a rather short Pd–Zn distance (2.794 Å), again in coincidence with the one calculated for ZnMeCl (2.710 Å). Finally, **CT-2** yields the *trans* product **3** by coordination of a *THF* molecule to the Zn reagent and subsequent Pd–Zn dissociation. The optimized geometries for the transition state and intermediates involved in this mechanism are shown in Fig. 4.14.

Unlike the reaction with ZnMeCl that was found to be endergonic, the overall Gibbs energy profile for the reaction with $ZnMe_2$ indicates that this reaction is exergonic by $2.1\,\text{kcal}\cdot\text{mol}^{-1}$. Thus, the reaction equilibrium, in this case, is displaced toward the products (**3** + $ZnMeCl(THF)_2$). As a consequence, the reverse reaction (i.e. retrotransmetalation from **3** to **1**) is slower than the forward reaction (i.e. transmetalation from **1** to **3**).

[12] For further details on this model, see the last section of Chap. 2.

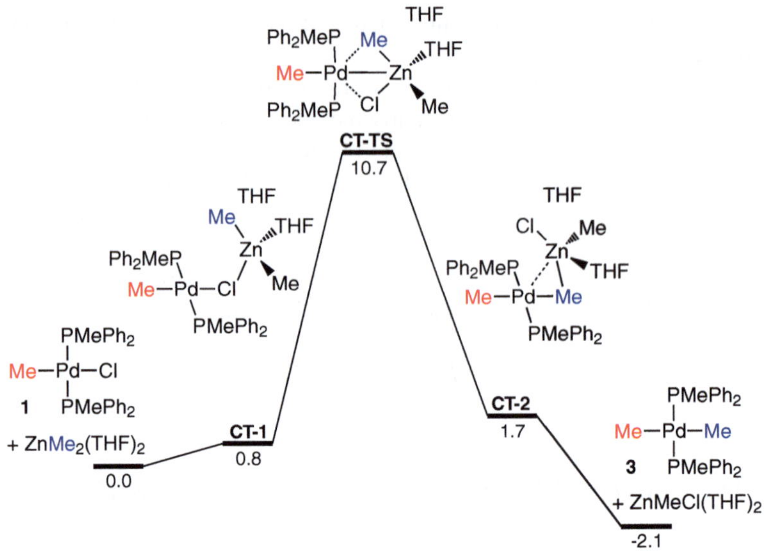

Fig. 4.13 Gibbs energy profile in *THF* (ΔG_{THF}, kcal · mol^{-1}) at 203 K for the transmetalation with $ZnMe_2$ leading to the *trans* product (**3**) via a concerted mechanism

Fig. 4.14 Optimized structures for the transition state and all the intermediates involved in the transmetalation with $ZnMe_2$ leading to the *trans* product (**3**) via a concerted mechanism. Phenyl rings of the phosphine ligands have been simplified and the second *THF* molecule of the organozinc reactant is not shown for clarity. Distances are shown in Å

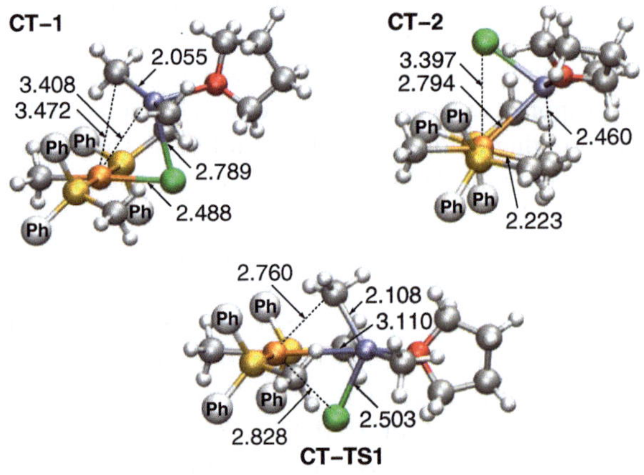

4.3.4 *The Concerted Mechanism to the* cis *Intermediate*

Following the same concerted mechanism than the one proposed in the study with ZnMeCl (Fig. 4.7), the transmetalation of **1** with $ZnMe_2$ leading to the *cis* product **2** was computed. The Gibbs energy profile obtained for the reaction through that mechanism is shown in Fig. 4.15.

As the other concerted pathways, this mechanism starts with the dissociation of one *THF* molecule from the reactant $ZnMe_2(THF)_2$ and the concomitant interaction

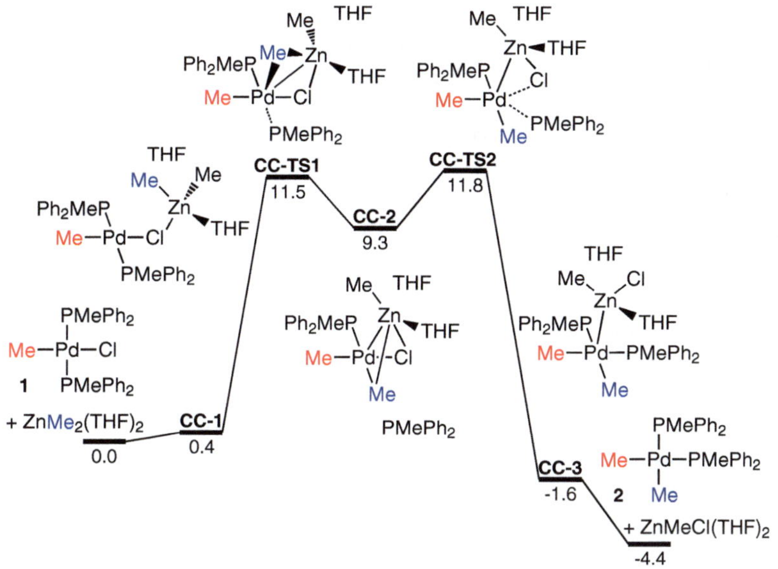

Fig. 4.15 Gibbs energy profile in *THF* (ΔG_{THF}, kcal · mol^{-1}) at 203 K for the transmetalation with $ZnMe_2$ leading to the *cis* product (**2**) via a concerted mechanism

between the Cl and Zn atoms. The resulting intermediate from this first step is **CC-1**, which lies only 0.4 kcal · mol^{-1} above the separated reactants (**1** + $ZnMe_2(THF)_2$). Then, **CC-1** undergoes phosphine by Me substitution followed by phosphine by chloride substitution. Unlike the reaction with ZnMeCl, where that double substitution was found to take place simultaneously (**CC-TS1**, Fig. 4.7), with $ZnMe_2$ it occurs in two separated steps. More specifically, the phosphine by Me substitution occurs through the transition state **CC-TS1** (11.5 kcal · mol^{-1}), while the following phosphine by chloride substitution takes place via **CC-TS2** (11.8 kcal · mol^{-1}). Interestingly, both transition states and the intermediate that connects them (**CC-2**) have similar Gibbs energies, which indicates that once the first substitution occurs, the second one is very easy (it requires only 2.5 kcal · mol^{-1}). Furthermore, all these species feature a four-membered ring formed between Pd, Cl, Zn, and Me atoms (Fig. 4.16) and fairly short Pd–Zn distances (the largest distance is 2.707 Å, in **CC-TS1**).

The second substitution through **CC-TS2** gives rise to the last intermediate **CC-3**, where the two Me groups are in *cis* disposition and the short Pd–Zn distance is maintained (2.700 Å). Finally, this intermediate yields the final product **2** by the coordination of the *THF* molecule to the Zn reagent and the subsequent Pd–Zn dissociation. The optimized geometries for all the transition states and intermediates that take part in this mechanism are displayed in Fig. 4.16.

Overall, the Gibbs energy profile depicted in Fig. 4.15 shows that transmetalation of **1** to the *cis* product **2** is exergonic by 4.4 kcal · mol^{-1}. Thus, the reaction equilibrium is displaced towards products (**2** + $ZnMeCl(THF)_2$), which reflects in a faster

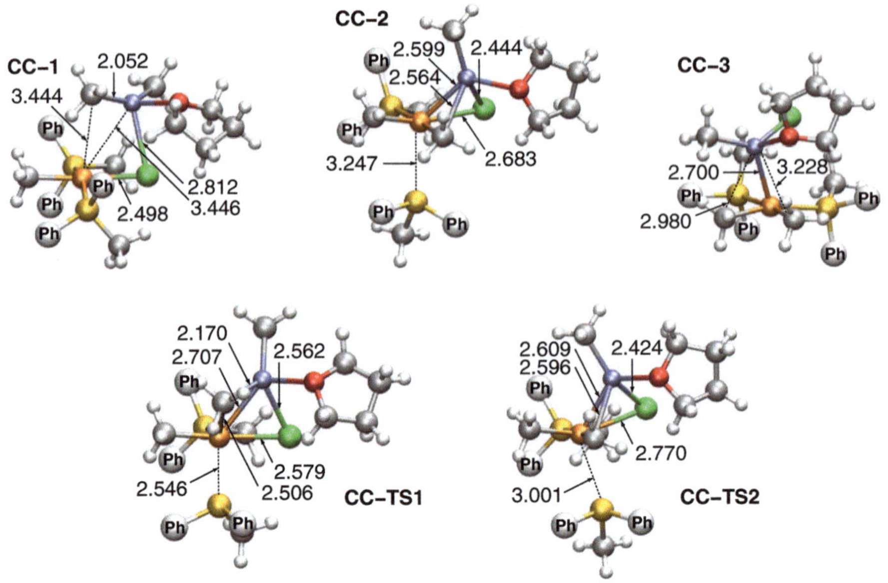

Fig. 4.16 Optimized structures for all the intermediates and transition states involved in the transmetalation with $ZnMe_2$ leading to the *cis* product (**2**) via a concerted mechanism. Phenyl rings of the phosphine ligands have been simplified and the second *THF* molecule of the organozinc reactant is not shown for clarity. Distances are shown in Å

transmetalation compared to retrotransmetalation. The highest energy barrier within this overall reaction profile corresponds to the chloride by phosphine substitution via **CC-TS2**, with an energy of 11.8 kcal · mol^{-1}.

4.3.5 The Ionic Mechanism to the Trans *Intermediate (L = THF)*

According to the experimental results, a consistent explanation for the unusually fast transmetalation observed with $ZnMe_2$ would be the formation of the non-observable intermediate *trans*-$[Pd(Me)(PMePh_2)_2(THF)]^+$ (**4**$^+$), from which the reaction would take place at a higher rate. Thus, in order to evaluate this proposal, a plausible transmetalation mechanism involving this cationic intermediate **4**$^+$ was theoretically investigated. The analyzed mechanism is reminiscent of the concerted mechanism leading to the *cis* product **2**, where the transmetalation reaction between Pd an Zn centers takes place in two steps (Fig. 4.15). In this case, however, the Me and Cl ligands are transferred in reversed order; that is, first the Cl atom is transferred from Pd to Zn, and afterwards, the Me group from Zn to Pd. On the other hand, this mechanism involves the formation of charged species and for this reason we named it

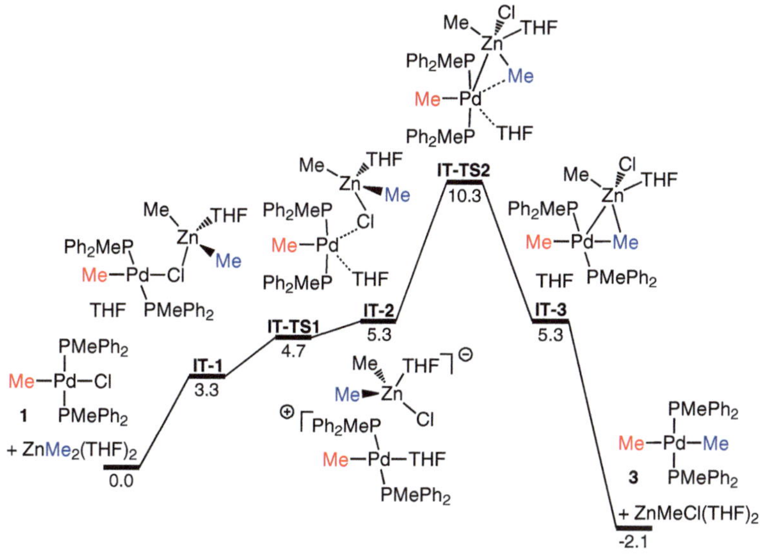

Fig. 4.17 Gibbs energy profile in *THF* (ΔG_{THF}, kcal · mol^{-1}) at 203 K for the transmetalation with $ZnMe_2$ leading to the *trans* product (**3**) via the ionic mechanism with *L* = *THF*

ionic mechanism. The calculated Gibbs energy profile for the transmetalation reaction through this mechanism is depicted in Fig. 4.17.

The first step in this ionic mechanism is the same than for the concerted mechanisms investigated so far. It entails the dissociation of a *THF* molecule from $ZnMe_2(THF)_2$ and the concomitant bridging chloride coordination to Zn. The intermediate that results from this step is **IT-1**,[13] which undergoes chloride abstraction assisted by an external coordinating ligand (*L*) and the Zn reagent. In this case, since *THF* is the solvent and furthermore has a moderate coordinating ability, it can act as external ligand and assist the chloride abstraction through the transition state **IT-TS1**[14] affording the unstable intermediate **IT-2** (corresponding to $\mathbf{4}^+$ in the experimental nomenclature). Importantly, the relative position of this last intermediate in the Gibbs energy profile is 5.3 kcal · mol^{-1} above the starting reactants, which makes its observation by ^{31}P NMR very unlikely.

Once **IT-2** is formed, the *THF* molecule coordinated to the cationic Pd complex *trans*-$[Pd(Me)(PMePh_2)_2(THF)]^+$ is replaced by a Me group of the counter-ion $[ZnMe_2Cl(THF)]^-$ with a relative energy barrier of only 5.0 kcal · mol^{-1}. The transition state involved in this substitution is **IT-TS2** and results in the last intermediate

[13] Note that this intermediate differs from their analogous **CT-1** and **CC-1** in that the dissociated *THF* molecule is explicitly included.

[14] This transition state appears 0.6 kcal · mol^{-1} below the following intermediate **IT-2**, which suggests that **IT-TS1** might not be a transition state in the Gibbs energy surface. In any case, the proximity in energy of **IT-TS1** and **IT-2** indicates that the backward process to generate the starting reactants is very easy.

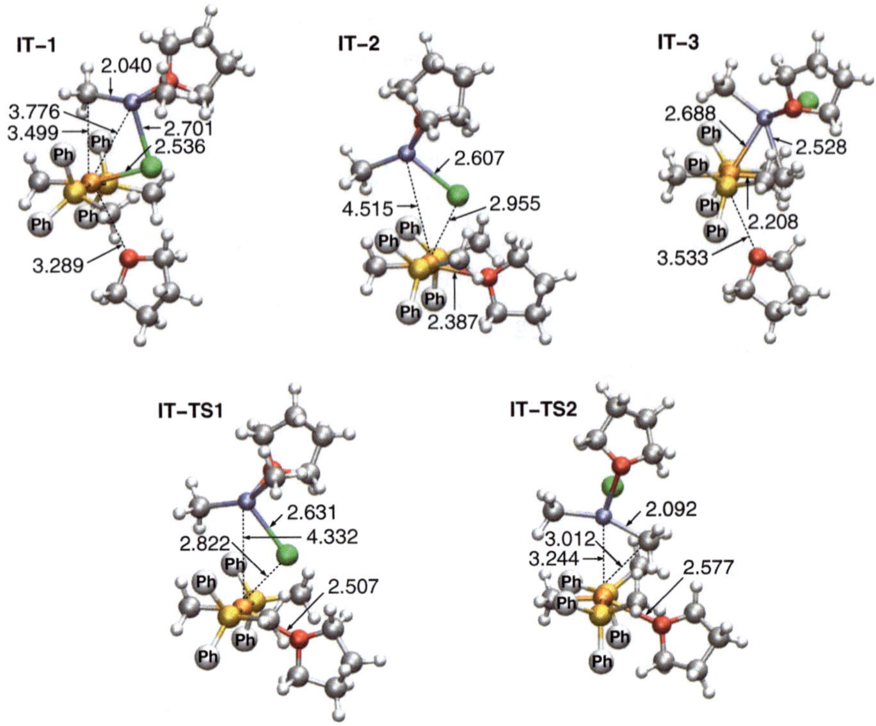

Fig. 4.18 Optimized structures for all the intermediates and transition states involved in the transmetalation with $ZnMe_2$ leading to the *trans* product (**3**) via the ionic mechanism with *L* = *THF*. Phenyl rings of the phosphine ligands have been simplified and the second *THF* molecule of the organozinc reactant is not shown for clarity. Distances are shown in Å

IT-3. Finally, this last intermediate evolves to the *trans* product **3** by the coordination of a *THF* molecule to the Zn reactant and the dissociation of the Pd–Zn complex. The optimized geometries for the transition states and intermediates involved in this ionic mechanism are shown in Fig. 4.18.

Overall, the calculated Gibbs energy profile depicted in Fig. 4.17 shows that the highest energy barrier within this ionic mechanism corresponds to the transfer of the Me group from Zn to Pd through **IT-TS2** (10.3 kcal · mol^{-1}). This energy barrier is even lower than that found for the concerted mechanism leading to **3** (Fig. 4.14), which consistently supports the proposal of an alternative transmetalation mechanism involving the cationic intermediate $\mathbf{4}^{+}$ as a plausible explanation for the unusual fast transmetalation observed.

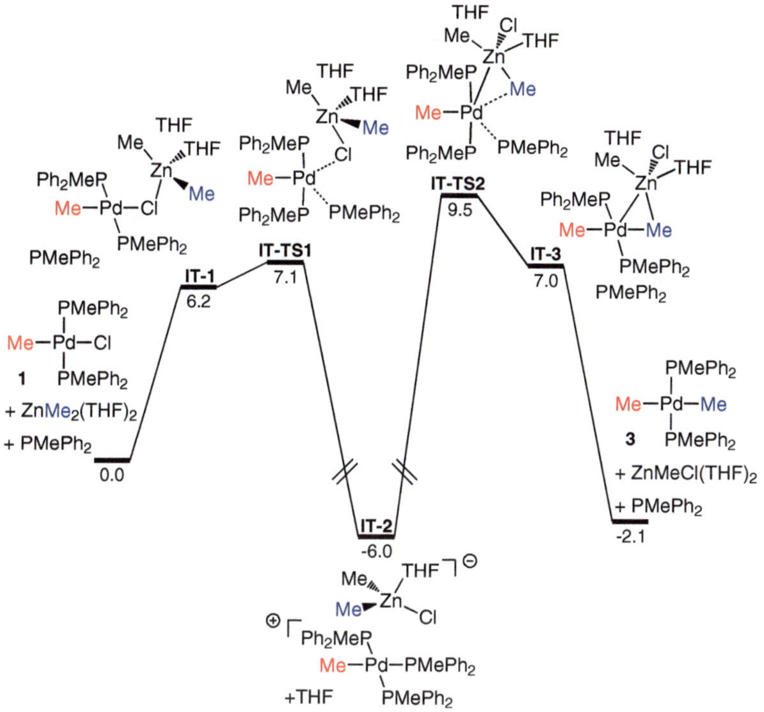

Fig. 4.19 Gibbs energy profile in *THF* (ΔG_{THF}, kcal · mol^{-1}) at 203 K for the transmetalation with $ZnMe_2$ leading to the *trans* product (**3**) via the ionic mechanism with $L = PMePh_2$

4.3.6 The Ionic Mechanism to the Trans *Intermediate ($L = PMePh_2$)*

Finally, in order to unravel the decrease of the reaction rates observed for the transmetalation process with added phosphine and further support the observation of the cationic species $[Pd(Me)(PMePh_2)_3]^+$ (**5**$^+$), a transmetalation pathway through this species was computed. This pathway involves the same steps than the ionic mechanism with $L = THF$ (Fig. 4.17), but entails very different energies (Fig. 4.19). Probably the most noteworthy difference between these two mechanisms is the different stability of intermediates **IT-2** corresponding to the cationic species **4**$^+$ (i.e. $L = THF$) and **5**$^+$ (i.e. $L = PMePh_2$). According to calculations, the former has a relative energy of 5.3 kcal · mol^{-1} above separated reactants, whereas the latter lies 6.0 kcal · mol^{-1} below separate reactants, which accounts for its observation by ^{31}P NMR. This lowering in the energy of **IT-2**, however, makes the energy barrier for the following step (**IT-TS2**) increasing to 15.5 kcal · mol^{-1}. The optimized geometries for all the transition states and intermediates implicated in this ionic mechanism are shown in Fig. 4.20.

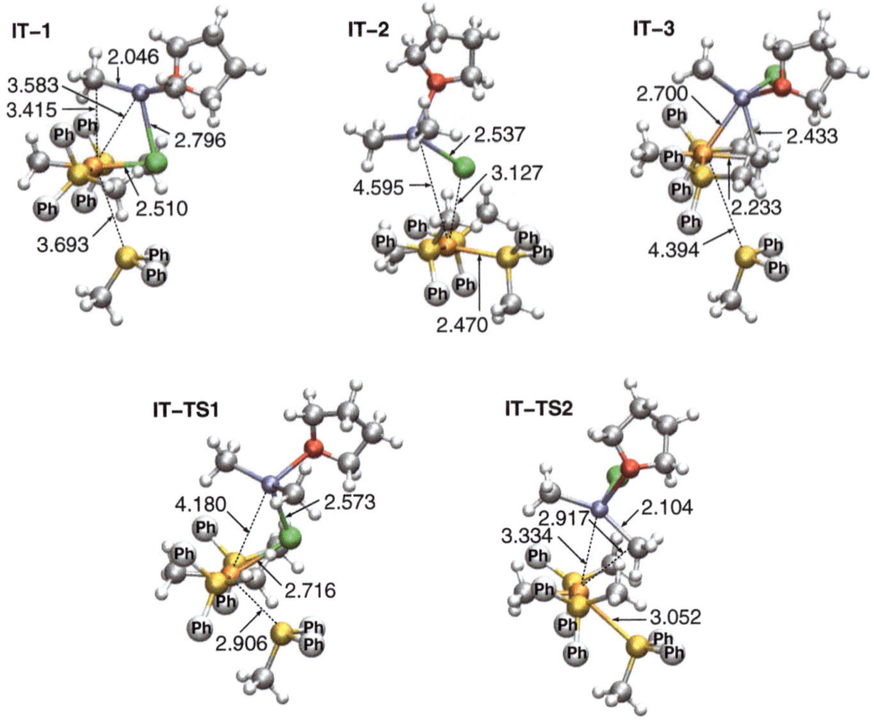

Fig. 4.20 Optimized structures for all the intermediates and transition states involved in the transmetalation with $ZnMe_2$ leading to the *trans* product (**3**) via the ionic mechanism with $L = PMePh_2$. Phenyl rings of the phosphine ligands have been simplified and the second *THF* molecule of the organozinc reactant is not shown for clarity. Distances are shown in Å

In summary, the computed Gibbs energy profile for the transmetalation through this ionic mechanism with $L = PMePh_2$ (Fig. 4.19) reveals that the very low energy of **IT-2** makes the observation of this species by ^{31}P NMR to be very likely. Similarly to the mechanism with $L = THF$, the highest energy barrier corresponds to the Me by phosphine substitution via **IT-TS2** ($15.5\,kcal \cdot mol^{-1}$). This energy is more than $3\,kcal \cdot mol^{-1}$ higher than those calculated for the concerted mechanisms and the ionic mechanism with $L = THF$, which in terms of reaction rates means more than 3 orders of magnitude slower. Therefore, this pathway is clearly the most disfavored one for the transmetalation of **1** with $ZnMe_2$.

4.3.7 Comparing Theoretical and Experimental Results

After having summarized the experimental and theoretical results obtained for the transmetalation of **1** with $ZnMe_2$, the only thing that is left is to compare them

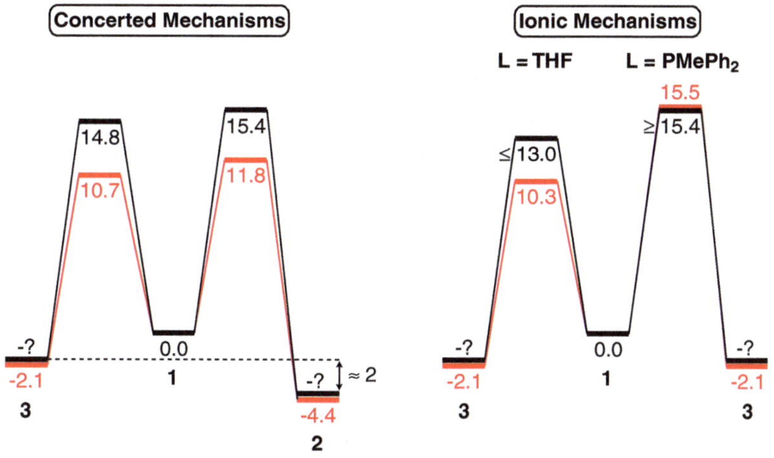

Fig. 4.21 Comparison of the Gibbs energy values at 203 K ($\Delta G_{223\,K}$, kcal · mol^{-1}) for the transmetalation reaction of **1** with $ZnMe_2$. Zn species are not shown for clarity. Experimental values are shown in *black*, whereas calculated values are shown in *red*

and draw the final conclusions of this mechanistic study. With this purpose, the Gibbs energies at 203 K obtained in the experimental[15] and theoretical studies on this reaction have been summarized in the simplified reaction profiles depicted in Fig. 4.21.

At first glance, from Fig. 4.21 it can be observed that calculations reproduce qualitatively all the experimental observations. In particular, from the thermodynamic perspective, the computed Gibbs energies confirm that transmetalation reactions leading to the *trans* and *cis* products (**3** and **2**, respectively) are exergonic, in coincidence with experiments. Furthermore, the computed order of stability of the reagents and products in equilibrium is the same than that observed in the experiments: **2** > **3** > **1**. On the other hand, from the kinetic point of view, the calculated Gibbs energy barriers for transmetalation through the different analyzed mechanisms decrease in the order: ionic to *trans* ($L = PMePh_2$) > concerted to *cis* > concerted to *trans* > ionic to *trans* ($L = THF$), which consistently accounts for the observation of the unusually fast transmetalation of **1** to the kinetic product **3** and the decelerating effect of the added phosphine (Fig. 4.11).

The quantitative comparison of experimental and theoretical Gibbs energy barriers can be only strictly made for the concerted mechanisms, since only the lower (for $\mathbf{5}^+$) and higher (for $\mathbf{4}^+$) limits for the ionic mechanisms could be estimated experimentally. Thus, according to the computed barriers for the concerted mechanisms

[15] These values were calculated using the reaction rates derived from the kinetic model *a* adopted in the experiments. Note that the experimental values for the concerted mechanisms are the quantitative kinetic results, whereas for the ionic mechanisms the values plotted correspond to the estimated maximum (i.e. $L = THF$) or minimum (i.e. $L = PMePh_2$) values compatible with the experimental observations.

it seems that they are underestimated by about 4 kcal · mol^{-1}. Nevertheless, calculations nicely reproduce the energy difference between the concerted mechanisms (0.6 kcal · mol^{-1} in the experiments, and 1.1 kcal · mol^{-1} in the theoretical values). Similarly, the relative stability of products **2** and **3** could not be quantified experimentally either, but in the study with ZnMeCl we demonstrated that **2** is about 2.0 kcal · mol^{-1} more stable than **3**. Hence, a calculated difference of 2.3 kcal · mol^{-1}, in this case, can be considered a very satisfactory fit with the experiment.

4.3.8 Conclusions

Following our research interest in the Negishi reaction, the transmetalation reaction between the complex *trans*-$[PdMeCl(PMePh_2)_2]$ (**1**) and $ZnMe_2$ was investigated both experimentally and theoretically. The results derived from this study showed that all the reactions with $ZnMe_2$ are faster than with ZnMeCl. Similarly to ZnMeCl, the transmetalation with $ZnMe_2$ was found to afford the *trans* product (**3**) at a higher rate than the *cis* product (**2**). However, the **3** to **2** isomerization, necessary for the success of the coupling, resulted to be slower than with ZnMeCl. This is due to the slower retrotransmetalation reactions with $ZnMe_2$.

Probably the most interesting observation of this work was that, besides the expected concerted pathways previously proposed for ZnMeCl, alternative ionic mechanisms could also be operative. These ionic pathways involve cationic intermediates of the type $[Pd(Me)(PMePh_2)_2(L)]^+$ $[ZnMe_2Cl(THF)]^-$ ($L = THF$, $PMePh_2$). In particular, the ionic mechanism with $L = THF$ is faster than the concerted mechanisms, while this mechanism with $L = PMePh_2$ is much more slower. In fact, the observation of the complex $[Pd(Me)(PMePh_2)_3]^+$ (**5**$^+$) suggests that the formation of this complex becomes a trap of the Pd catalyst. This observation consistently accounts for the decelerating effect of the addition of $PMePh_2$ to the reaction of **1** with $ZnMe_2$. Interestingly, the fact that no accelerating effect was detected in the study with ZnMeCl can be attributed to the unlikely formation of the cationic species $[Pd(Me)(PMePh_2)_2(THF)]^+$ (**4**$^+$), since the presence of a large number of terminal chloro donor atoms will prevent *THF* from coordinating Pd.

All the results obtained in this combined experimental and theoretical study are summarized in the catalytic cycle depicted in Fig. 4.22.

Overall, this study provided a proof of existence, operation and effects of competitive transmetalation pathways, some of which had not been invoked before, but that should be considered from now on when planning or discussing Negishi syntheses. For instance, the oxidative addition products $[Pd(R)(L)_2(solvent)]^+$ X^- generated by the addition of R–X to $[PdL_2]$, in the case of weakly coordinating anions X, are very likely to promote ionic mechanisms. Conversely, a less donating R group on Pd, instead of Me, is expected to be somewhat less prone to produce ionic species, whereas heavier halogens will promote ionic Pd species better than Cl. As far as the solvent is concerned, the use of solvents more coordinating than *THF* will also favor ionic mechanisms.

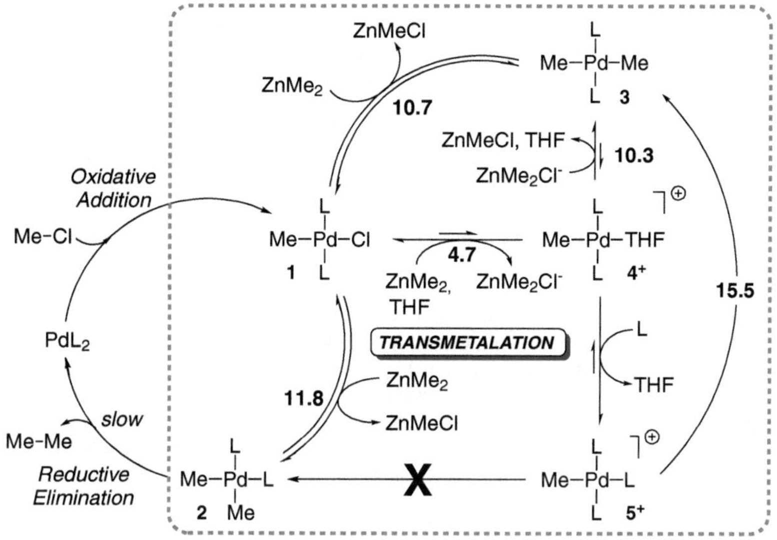

Fig. 4.22 Simplified reaction mechanism for the Pd-catalyzed Negishi cross-coupling reaction with $ZnMe_2$ ($L = PMePh_2$)

Focusing on the computational part, this study also showed that matching experimental relative Gibbs energies in solution with theoretical calculations is a harsh work even when working with the real system and a good methodology. Overall, calculated Gibbs energies reproduce better the relative stabilities among the different intermediates than their relative energy barriers associated to these processes. However, calculations revealed that the computation of energy differences between reaction mechanisms are very useful to compare their feasibility.

References

1. de Meijere, A., Diedrich, F. (eds.): Metal-Catalyzed Cross-Coupling Reactions (2nd edn). Wiley-VCH, Weinheim (2004)
2. Hartwig, J.: Organotransition Metal Chemistry: from Bonding to Catalysis. University Science Books, California (2010)
3. Negishi, E.-I., King, A.O., Okukado, N.: J. Org. Chem. **1977**, 42 (1821)
4. Phapale, V.B., Cárdenas, D.J.: Chem. Soc. Rev. **38**, 1598 (2009)
5. Negishi, E.-I.: Acc. Chem. Res. **15**, 340 (1982)
6. Knochel, P., Singer, R.D.: Chem. Rev. **93**, 2117 (1993)
7. Rouhi, A.M.: Chem. Eng. News **82**, 49 (2004)
8. King, A.O., Yasuda, N.: Organometallics in Process Chemistry Vol. 6 of Topics in Organometallic Chemistry. Springer, Berlin (2004)
9. Hirashima, S., Aoyagi, S., Kibayashi, C.: J. Am. Chem. Soc. **121**, 9873 (1999)
10. Wipf, P., Lim, S.: J. Am. Chem. Soc. **117**, 558 (1995)
11. Negishi, E.-I.: Bull. Chem. Soc. Jpn. **80**, 233 (2007)

12. Espinet, P., Echavarren, A.M.: Angew. Chem. Int. Ed. **43**, 4704 (2004)
13. Casado, A.L., Espinet, P.: J. Am. Chem. Soc. **120**, 8978 (1998)
14. Nova, A., Ujaque, G., Maseras, F., Lledós, A., Espinet, P.: J. Am. Chem. Soc. **128**, 14571 (2006)
15. Álvarez, R., Faza, O.N., López, C.S., de Lera, A.R.: Org. Lett. **8**, 35 (2006)
16. Álvarez, R., Pérez, M., Faza, O.N., de Lera, A.R.: Organometallics **27**, 3378 (2008)
17. Álvarez, R., Faza, O.N.: Cárdenas. D. J. Adv. Synth. Catal. **349**, 887 (2007)
18. Braga, A.A.C., Ujaque, G., Maseras, F.: Organometallics **25**, 3647 (2006)
19. Braga, A.A.C., Morgon, N.H., Ujaque, G., Lledós, A., Maseras, F.: J. Organomet. Chem. **691**, 4459 (2006)
20. Sicre, C., Braga, A.A.C., Maseras, F., Cid, M.M.: Tetrahedron **64**, 7437 (2008)
21. Carrow, B.P., Hartwig, J.F.: J. Am. Chem. Soc. **133**, 2116 (2011)
22. Urkalan, K., Sigman, M.S.: J. Am. Chem. Soc. **131**, 18042 (2009)
23. Howell, G.P., Minnaard, A.J., Feringa, B.L.: Org. Biomol. Chem. **4**, 1278 (2006)
24. Marshall, J.A.: Chem. Rev. **100**, 3163 (2000)
25. Jin, L., Zhao, Y., Zhu, L., Zhang, H., Lei, A.: Adv. Synth. Catal. **351**, 630 (2009)
26. Chen, M., Zheng, X., Li, W., He, J., Lei, A.: J. Am. Chem. Soc. **132**, 4101 (2010)
27. Casares, J.A., Espinet, P., Fuentes, B., Salas, G.: J. Am. Chem. Soc. **129**, 3508 (2007)
28. Van Asselt, R., Elsevier, C.J.: Organometallics **13**, 1972 (1994)
29. Liu, Q., Lan, Y., Liu, J., Li, G., Wu, Y.D., Lei, A.: J. Am. Chem. Soc. **131**, 10201 (2009)
30. Chass, G.A., O'Brien, C.J., Hadei, N., Kantchev, E.A.B., Mu, W.-H., Fang, D.-C., Hopkinson, A.C., Csizmadia, I.G., Organ, M.G.: Chem. Eur. J. **15**, 4281 (2009)
31. Ribagnac, P., Blug, M., Villa-Uribe, J., Le Goff, X.-F., Gosmini, C., Mézailles, N.: Chem. Eur. J. **17**, 14389 (2011)
32. Mendes, P.: Comput. Appl. Biosci. **9**, 563 (1993)
33. Zhao, Y., Truhlar, D.G.: Theor. Chem. Acc. **120**, 215 (2008)
34. Frisch, M.J., Trucks, G.W., Schlegel, H.B., Scuseria, G.E., Robb, M.A., Cheeseman, J.R., Scalmani, G., Barone, V., Mennucci, B., Petersson, G.A., Nakatsuji, H., Caricato, M., Li, X., Hratchian, H.P., Izmaylov, A.F., Bloino, J., Zheng, G., Sonnenberg, J.L., Hada, M., Ehara, M., Toyota, K., Fukuda, R., Hasegawa, J., Ishida, M., Nakajima, T., Honda, Y., Kitao, O., Nakai, H., Vreven, T., Montgomery, Jr., J. A., Peralta, J.E., Ogliaro, F., Bearpark, M., Heyd, J.J., Brothers, E., Kudin, K.N., Staroverov, V.N., Kobayashi, R., Normand, J., Raghavachari, K., Rendell, A., Burant, J.C., Iyengar, S.S., Tomasi, J., Cossi, M., Rega, N., Millam, J.M., Klene, M., Knox, J.E., Cross, J.B., Bakken, V., Adamo, C., Jaramillo, J., Gomperts, R., Stratmann, R.E., Yazyev, O., Austin, A.J., Cammi, R., Pomelli, C., Ochterski, J.W., Martin, R.L., Morokuma, K., Zakrzewski, V.G., Voth, G.A., Salvador, P., Dannenberg, J.J., Dapprich, S., Daniels, A.D., Farkas, Ö., Foresman, J.B., Ortiz, J.V., Cioslowski, J., Fox, D.J.: Gaussian 09, Revision A.1, Gaussian, Inc., Wallingford, (2009)
35. Weidenbruch, M., Herrndorf, M., Schäfer, A., Pohl, S., Saak, W.: J. Organomet. Chem. **361**, 139 (1989)
36. Hay, P.J., Wadt, W.R.: J. Chem. Phys. **82**, pp. 270, 299 (1985)
37. Godbout, N., Salahub, D.R., Andzelm, J., Wimmer, E.: Can. J. Chem. **70**, 560 (1992)
38. Schäfer, A., Huber, C., Ahlrichs, R.: J. Chem. Phys. **100**, 5829 (1994)
39. Miertus, S., Scrocco, E., Tomasi, J.: J. Chem. Phys. **55**, 117 (1981)
40. Álvarez, R., de Lera, A.R., Aurrecoechea, J.M., Durana, A.: Organometallics **26**, 2799 (2007)
41. Fuentes, B., García-Melchor, M., Lledós, A., Maseras, F., Casares, J.A., Ujaque, G., Espinet, P.: Chem. Eur. J. **16**, 8596 (2010)
42. Spielvogel, D.J., Davis, W.M., Buchwald, S.L.: Organometallics **21**, 3833 (2002)
43. Spielvogel, D.J., Buchwald, S.L.: J. Am. Chem. Soc. **124**, 3500 (2002)
44. Anderson, T.J., Vicic, D.A.: Organometallics **23**, 623 (2004)
45. García-Melchor, M., Fuentes, B., Lledós, A., Casares, J.A., Ujaque, G., Espinet, P.: J. Am. Chem. Soc. **133**, 13519 (2011)
46. Marenich, A.V., Cramer, C.J., Truhlar, D.G.: J. Phys. Chem. B **113**, 6378 (2009)

Chapter 5
The Cu-Free Sonogashira Reaction Mechanism

Important scientific discoveries go through three phases: first they are completely ignored, then they are violently attacked, and finally they are brushed aside as well known
Max Tegmark, Nature, **2007**, *448*, 23–24

5.1 Introduction

The palladium-catalyzed Sonogashira reaction is one of the most important and widely used methods for preparing arylacetylenes and conjugated enynes [1–6], which are precursors for natural products, pharmaceuticals, and materials with specialized optical and electronic properties [7–12]. The two studies first reported on this field were independently published in 1975 by Heck's group [13] and Cassar [14]. The Heck and co-workers' procedure was based on the Mizoroki-Heck reaction [15, 16], namely the palladium-catalyzed arylation or alkenylation of alkenes, employing a phosphane-palladium complex as catalyst and triethylamine or piperidine both as base and solvent. On the other hand, Cassar's procedure also involved a phosphane-palladium complex as catalyst but, in this case, in combination with sodium methoxide as base and *DMF* as solvent. Both methods, however, had the disadvantage that generally required high temperatures (up to 100 °C). Afterwards but in the same year, Sonogashira and Hagihara reported that addition of a catalytic amount of CuI greatly accelerates the alkynylation reaction, thus enabling the performance of this reaction even at room temperature [17]. As a result, this Sonogashira-Hagihara reaction, commonly known simply as Sonogashira reaction, became the most popular

The results presented in this chapter have been published in the following article:
Article III: García-Melchor, M.; Pacheco, M. C.; Náijera, C.; Lledós, A.; Ujaque, G. *ACS Catal.* **2012**, *2*, 135–144.

M. G. Melchor, *A Theoretical Study of Pd-Catalyzed C–C Cross-Coupling Reactions*, Springer Theses, DOI: 10.1007/978-3-319-01490-6_5,

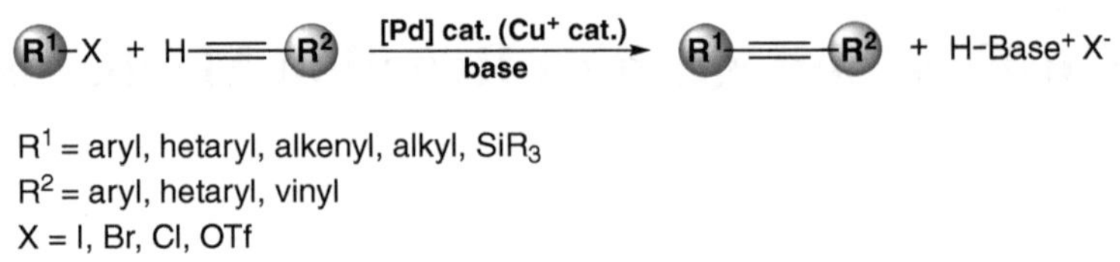

Scheme 5.1 General scheme for Pd-catalyzed Sonogashira reaction

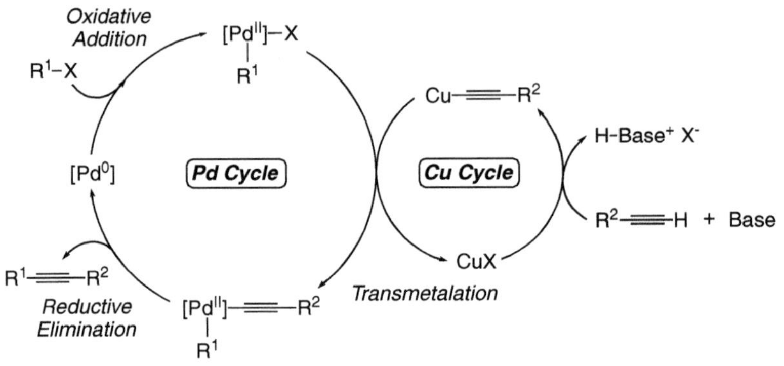

Fig. 5.1 General reaction mechanism for the Pd-catalyzed Sonogashira reaction

procedure for the alkynylation of aryl or alkenyl halides replacing all the other related protocols. Thereafter, the general Sonogashira protocol for the coupling of terminal alkynes with aryl or alkenyl halides (or triflates) usually involves an organic solvent, a Pd(0)/Cu(I) catalytic system, and at least a stoichiometric amount of a base (Scheme 5.1).

Although the exact mechanism for the Sonogashira reaction is still at this moment not fully understood, it is generally supposed to occur via two coupled catalytic cycles (Fig. 5.1): a "*Pd cycle*" and a "*Cu cycle*". The former is the classical from C–C cross-coupling reactions (described in Chap. 1) and starts up with the oxidative addition of the organic electrophile R^1–X to the catalytically active species [Pd(0)] to yield a [Pd(II)] complex with two new metal-ligand bonds. At this point, the next step in this cycle connects with the cycle of the copper co-catalyst (i.e. the Cu cycle). Unlike the Pd cycle, which is quite well understood, the Cu cycle is poorly known. In particular, in this cycle, the base is believed to assist the copper acetylide formation through a π-alkyne copper complex, which would make the alkyne terminal proton more acidic. Once the copper acetylide is formed, then it would undergo transmetalation reaction with the [Pd(II)] complex generated in the Pd cycle resulting in the complex *cis*-[Pd(R^1)(acetylide)], which affords the final product and the regeneration of the catalytic species through a common reductive elimination.

Thus, in principle, the presence of a copper(I) salt as co-catalyst in the typical Sonogashira reaction facilitates the reaction by the in situ generation of copper acetylide. However, the addition of this copper(I) salt under the reaction conditions entails some drawbacks, [18, 19] mainly the induction of the so-called *Glaser-type* homocoupling yielding the corresponding diyne [15, 20]. Aimed at suppressing the formation of this by-product, many efforts have been devoted to develop reaction procedures working in the absence of copper salts [21–29]. All these copper-free strategies are commonly known as *copper-free Sonogashira reactions*. Unfortunately, this copper-free variant usually requires the use of an excess of amine (often even acting as solvent), which proves detrimental to the environmental and economical advantages of this methodology. In order to avoid that, several modifications of the original Sonogashira reaction have been also recently reported including amine-free, ligand-free and solvent-free conditions [30–40]. These modifications, however, have been based on assumptions about a hypothetical reaction mechanism since, as mentioned, very little is known about the exact mechanism of the Sonogashira reaction, specially for the copper-free variant.

As far as the reaction mechanism for the copper-free Sonogashira reaction is concerned, two different mechanisms have been proposed (Fig. 5.2): the *deprotonation* [22] and *carbopalladation* [13] mechanisms. Both mechanisms share the initial oxidative addition of the organohalide R^1–X to the [$PdL_2(0)$] complex giving the intermediate **1** and the subsequent ligand by alkyne substitution from this species, which results in the formation of complex **2**. At this point, the two reaction mechanisms differ in the next steps leading to the final coupled product. More specifically, in the case of the deprotonation mechanism (Fig. 5.2, left), the deprotonation of the alkyne and the coordination of the ligand L take place from **2** yielding a square planar Pd complex with the two organic groups in *cis* disposition, from which the coupled product is expelled by reductive elimination. Alternatively, in the carbopalladation mechanism (Fig. 5.2, right), complex **2** undergoes addition of the organic group R^1 to the terminal alkyne, followed by the coordination of the ligand L and subsequent base-mediated reductive elimination.

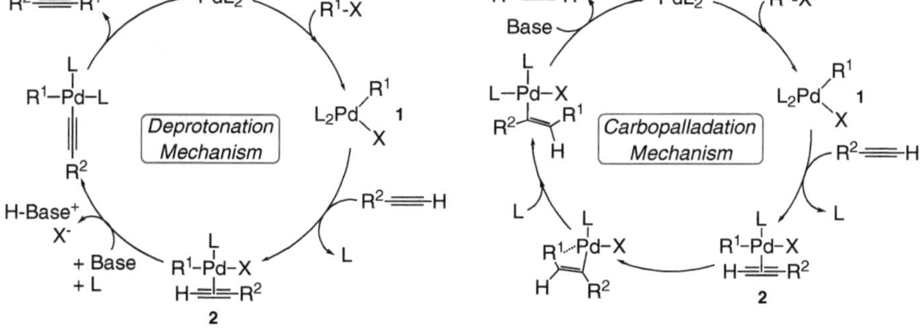

Fig. 5.2 Proposed reaction mechanisms for the copper-free Sonogashira reaction: deprotonation (*left*) and carbopalladation (*right*) mechanisms

Over the last years, these two mechanisms have been somewhat discussed in the literature, but the mechanism that operates in the copper-free Sonogashira reaction remains still unclear. In fact, up to date, the reported experimental and theoretical mechanistic studies on this process have been rather scarce. Among the experimental ones, one must mention the ones reported by Jutand et al. [41, 42] and Mårtensson et al. [43]. On the one hand, Jutand et al. by means of a thoughtful work shed light on the decelerating effect of alkynes in the oxidative addition step and suggested that amines might have multiple roles in the reaction mechanism. On the other hand, Mårtensson et al. demonstrated with a simple though clever experiment that the carbopalladation mechanism can be discarded, and proposed two alternatives routes for the deprotonation pathway. These two variants of the deprotonation mechanism, labeled by the authors as *cationic* and *anionic mechanisms*, only differ in the order in which the steps in this mechanism occur (Fig. 5.3). Thus, in the cationic mechanism (Fig. 5.3, left), the L-for-X ligand substitution in **2** takes place first giving rise to the cationic Pd complex *cis*-$[Pd(R^1)(alkyne)(L)_2]^+$, which undergoes deprotonation of the alkyne by an external base and subsequent reductive elimination. In contrast, in the anionic mechanism (Fig. 5.3, right), the deprotonation of the alkyne occurs first resulting in the anionic complex *cis*-$[Pd(R^1)(acetylide)(X)(L)]^-$, from which the L-for-X ligand substitution takes place followed by the reductive elimination step. According to Mårtensson et al., these cationic and anionic alternatives can be favored depending on the electronic nature of the substituents directly attached to the terminal alkynes. More specifically, the authors suggested that alkynes bearing electron withdrawing groups (*EWGs*) may favor the anionic mechanism, whereas alkynes bearing electron donating groups (*EDGs*) may favor the cationic pathway.

As regards to theoretical mechanistic studies on the copper-free variant, to the best of our knowledge, before we started our mechanistic study there was only the one reported by Chen et al. [44], whereas the work of Sikk et al. [45] appeared

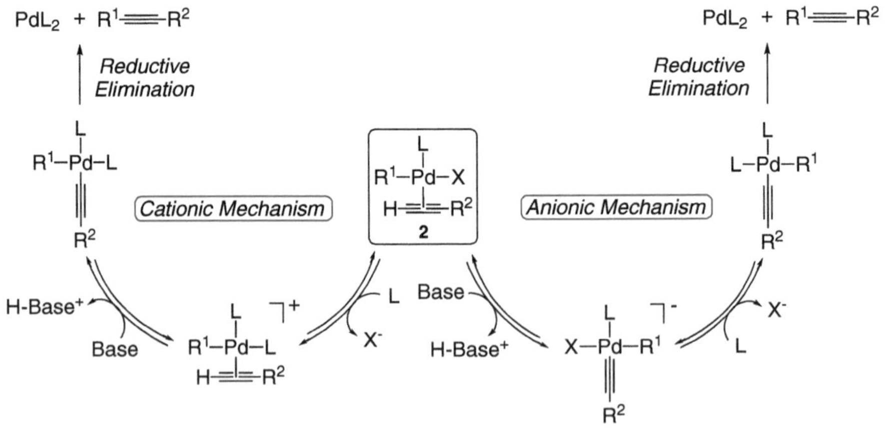

Fig. 5.3 Cationic (*left*) and anionic (*right*) alternatives for the deprotonation mechanism

during the revision process of our submitted article [46]. The former is on a model copper/base-free Sonogashira reaction in which the authors considered the halide (i.e. Br^-) as the species accepting the proton from the alkyne, while the latter is on a typical copper-free Sonogashira reaction but in which only a deprotonation-type mechanism was analyzed. Thus, in spite of these works, the dichotomy between the carbopalladation and deprotonation mechanisms, as well as additional alternatives, remained still unsolved.

In view of this picture, we decided to carry out a theoretical study in order to gain a deeper understanding of the reaction mechanism for this process. In particular, the purpose of our work was to weigh up the reaction mechanisms proposed in the literature and possible alternative pathways using a general model of the typical Pd-catalyzed copper-free Sonogashira reaction. To this aim, we computed the Gibbs energy profiles of the different mechanistic proposals by means of *DFT* calculations. Additionally, we also analyzed the effect of the electronic nature of the alkyne's substituents over all the studied reaction pathways. This feature, in particular, was investigated for a set of substituents combining theory and experiments. This work was carried out in collaboration with the experimental group of Prof. Nájera at the Universidad de Alicante.

5.2 Computational Details

All calculations in this study were performed at the *DFT* level, by means of the hybrid *B3LYP* [47, 48] functional and using the *Gaussian03* program [49]. Pd and I atoms were described using the Stuttgart-Dresden (*SDD*) effective core potential [50] for the inner electrons and its associated double-ζ basis set for the outer ones. Additionally, for these atoms f-polarization (exponent = 1.472) [51] and d-polarization (exponent = 0.289) [52] shells were added, respectively. In the case of I atoms diffuse functions were also added (exponent = 0.0308) [53]. For the C, P, H atoms and the N atoms the 6-31G(d,p) and the 6-31+G(d) basis sets were used, respectively. We chose such computational level because it has been widely employed in theoretical studies on related cross-coupling reactions providing good results [54–60].

All the structures of the reactants, intermediates and transition states were fully optimized without any symmetry constraint. Harmonic force constants were computed at the optimized geometries to characterize the stationary points as minima or saddle points. The latter were confirmed by having one imaginary frequency in the Hessian matrix and correlating the corresponding reactants and products. The entropic contributions were evaluated at a pressure of 382 atm following the argument in Martin's et al. [61] paper in order to model the changes in entropy for a condensed phase (i.e. dichloromethane).

Solvent effects (i.e. dichloromethane, $\varepsilon = 8.930$) were introduced through single point calculations at optimized gas-phase geometries for all the minima and transition states by means of a continuum method, the *PCM* approach [62] implemented in *Gaussian03*. Moreover, the default cavity model (i.e. *UA 0*) was modified by adding individual spheres to the hydrogen atoms directly linked to the alkyne and to the nitrogen atom of the pyrrolidine molecule, using the keyword *"SPHEREONH"*.

Unless otherwise stated, all the energies shown throughout this chapter correspond to Gibbs energies in dichloromethane (ΔG_{DCM}) at 298 K, obtained by employing the following scheme:

$$\Delta G_{DCM} = \Delta E_{DCM} + (\Delta G_{gas} - \Delta E_{gas})$$

5.3 Selection of the Model

In order to obtain a general overview on the process, we selected a set of model molecules for the analysis of the general reaction mechanism. Hence, phenylacetylene and iodobenzene species were selected as models for the coupling organic reactants, pyrrolidine as base, and dichloromethane (*DCM*) as solvent. On the other hand, for the additional analysis of the effect of the alkyne's substituents, several 4-substituted phenylacetylenes (R = H, CF_3, OMe, NMe_2) were considered (Scheme 5.2). As far as the catalyst is concerned, all the calculations were performed using $[Pd(PH_3)_2]$ as a model for the catalyst. Due to the increase of computational power, this model use to be considered too small, and catalysts with bigger phosphine ligands use to be calculated. Nevertheless, we chose this model in purpose for this study for the following reasons: *(i)* as our group have recently shown, conformational diversity induced by bulky phosphine ligands may introduce significant errors in the calculations of the energy profiles (with modifications in energy barriers higher than 10 kcal $\cdot$ mol^{-1} depending on the phosphines); [63] *(ii)* regarding the electronic properties of the phosphine, PH_3 can be considered as a neutral one; and *(iii)* in terms of computational requirements it is by far the less demanding one.

The main objective of our work was to map the potential energy surface by analyzing all the reasonable reaction profiles. Thus, in our opinion, this model covers most of the general features of the reaction system, apart from the steric effects that should be evaluated for each particular system.

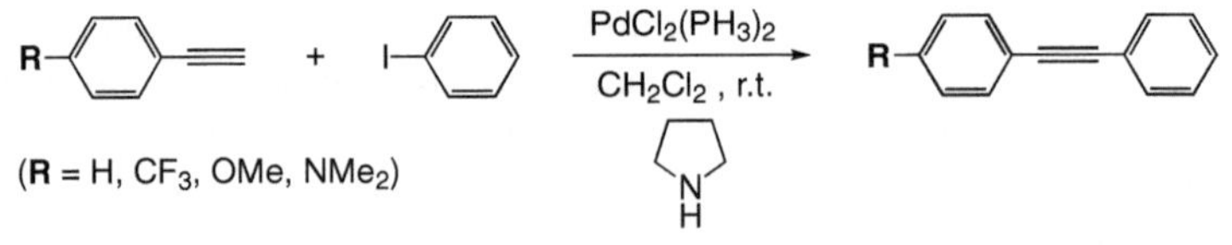

Scheme 5.2 Copper-free Sonogashira reaction between several 4-substituted phenylacetylenes and iodobenzene

5.4 The Oxidative Addition Step

The first step in all the proposed mechanisms for the copper-free Sonogashira reaction corresponds to the oxidative addition of the organic halide R–X to the starting [Pd(0)] complex. This step has been extensively studied (see Chap. 1) and is well known that in the case of organic iodides does not use to be rate-limiting. Even so, we decided to examine it for completeness. Hence, the oxidative addition of PhI to the complex [Pd(PH_3)$_2$] was computed. The optimized structures for this process are shown in Fig. 5.4.

As expected the calculated energy barrier for the oxidative addition reaction was rather low (17.0 kcal $\cdot$ mol^{-1}) and involves the concerted formation of the Pd–I and Pd–C bonds, and the cleavage of the C–I bond through a three-centered transition state (**OA-TS**). This transition state results in the oxidative addition product **OA-P**, which evolves to the more stable *trans* isomer through a cis-to-trans isomerization. This isomerization is known that may take place following different pathways, [54] but in any case it is an easy process [64]. Thus, we focused our further analysis on the proposed mechanisms starting from the *trans*-[Pd(Ph)(I)(PH_3)$_2$] (**1**) complex.

5.5 The Carbopalladation Mechanism

Despite Mårtensson et al. [43] experimentally showed that the carbopalladation mechanism can not be operative, in order to have a comprehensive mechanistic understanding of the reaction we decided to investigate this mechanism together with the deprotonation mechanism including their cationic and anionic alternatives (see below).

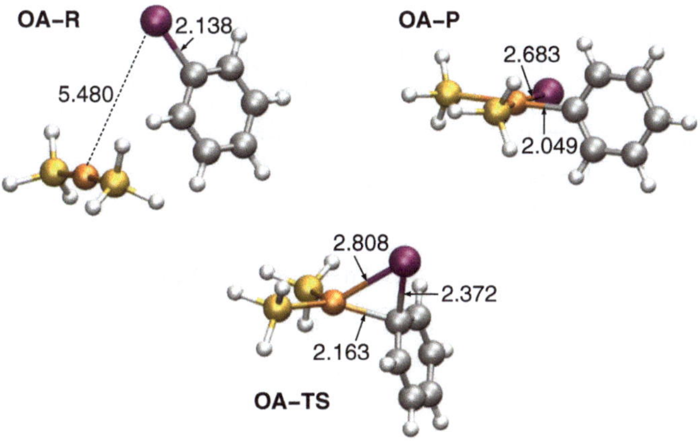

Fig. 5.4 Optimized structures for the reactant (**OA-R**), product (**OA-P**) and transition state (**OA-TS**) in the oxidative addition of PhI to [Pd(PH_3)$_2$]. Distances are shown in Å

The theoretical investigation of the copper-free Sonogashira reaction with phenylacetylene as a model substrate (R = H) through a carbopalladation mechanism afforded the reaction profile shown in Fig. 5.5.

As previously mentioned, the carbopalladation and the deprotonation mechanisms share the initial substitution of a phosphine ligand by the alkyne, which leads to the formation of a common intermediate (**2**). The calculation of this step via an associative substitution indicated that this is an endergonic process by 13.5 kcal · mol^{-1} with an energy barrier of 23.0 kcal · mol^{-1}. Once complex **2** is formed, the carbopalladation reaction occurs through the transition state **C-TS1** resulting in the intermediate **C-1** with a relative energy barrier of 14.2 kcal · mol^{-1}. Subsequently, this intermediate evolves with a low energy barrier (6.4 kcal · mol^{-1}) to the very stable intermediate **C-2** by coordination of a phosphine ligand via **C-TS2**. Finally, the alkenyl moiety in **C-2** is deprotonated by the external base through the transition state **C-TS3** yielding the final coupled product (**3**) and the regeneration of the catalytic species. This last step has the highest energy barrier in the overall energy profile (40.4 kcal · mol^{-1}), which can be attributed to the high stability of **C-2** and the difficult that entails the deprotonation of a double bond (high energy of **C-TS3**). The optimized structures for all the transition states involved in this mechanism are shown in Fig. 5.6.

Overall, the reaction is exergonic by 21.5 kcal · mol^{-1} but the carbopalladation mechanism has a very high energy barrier (40.4 kcal · mol^{-1}), which makes this

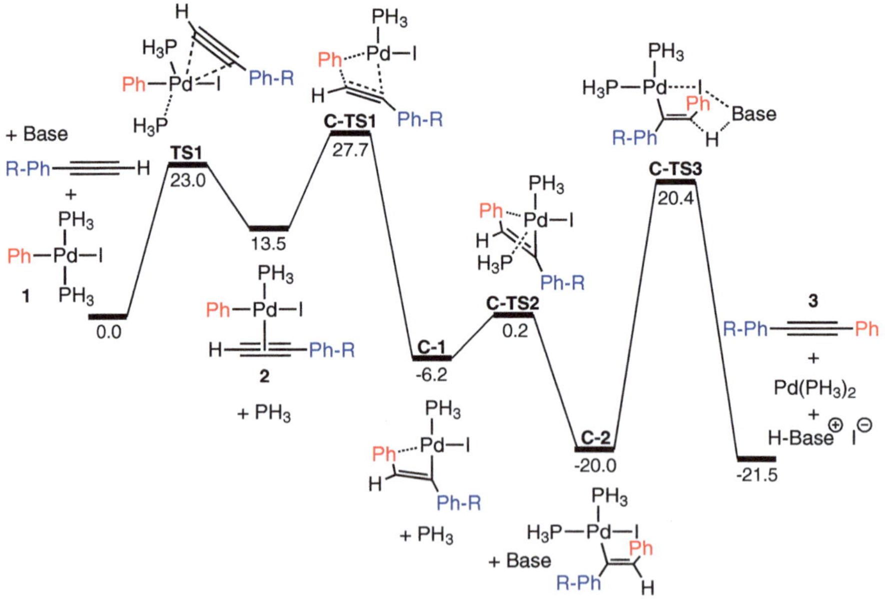

Fig. 5.5 Gibbs energy profile in *DCM* (ΔG_{DCM}, kcal · mol^{-1}) at 298 K for the carbopalladation mechanism with R = H, and Base = pyrrolidine

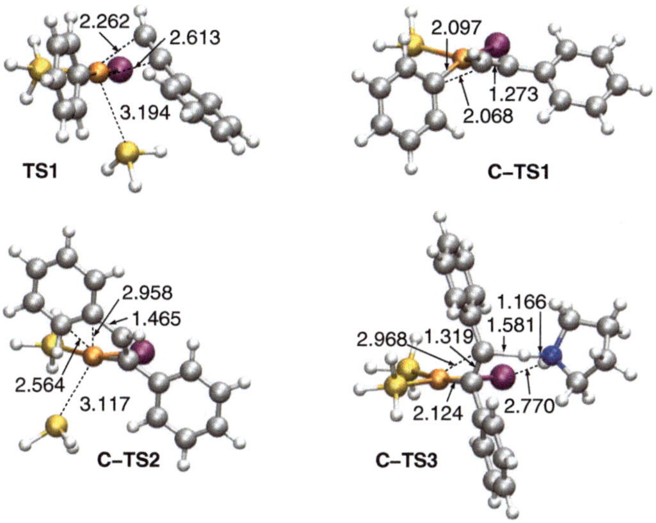

Fig. 5.6 Optimized structures for the transition states involved in the carbopalladation mechanism with phenylacetylene (R = H). Distances are shown in Å

mechanism very unlikely to be operating in the copper-free Sonogashira reaction. The occurrence of a very stable intermediate placed 20 kcal · mol^{-1} below reactants (i.e. **C-2**) and the necessity of overcoming a barrier of about 40 kcal · mol^{-1} (energy from **C-2** to **C-TS3**) are responsible for the unsuitability of the carbopalladation mechanism. This theoretical finding for phenylacetylene (R = H) agrees with the experimental observation of Mårtensson et al. [43] that a complex analogous to **C-2** synthesized through an alternative route does not afford the coupled product under the Sonogashira reaction conditions.

With the carbopalladation reaction pathway established for phenylacetylene (R = H) and for the sake of completeness, the effect of the alkyne R substituent on the overall carbopalladation mechanism was next examined. With this purpose, the Gibbs energy profiles for the Sonogashira reaction with several 4-substituted phenylacetylenes (R = CF_3, OMe, NMe_2) through a carbopalladation mechanism were computed (Table 5.1).

The results collected in Table 5.1 show that the influence of the substituent on the reaction energy is rather low: the reaction is highly exergonic with all the substituents by around 21 kcal · mol^{-1}. Moreover, the results also show that the most important effect of the different R groups is in the step common to all the different mechanisms, which is the coordination of the alkyne to the palladium complex **1** to yield complex **2**. The calculated energy barrier for this process (Table 5.1, **TS1**) decreases in the order: R = CF_3 > H > OMe > NMe_2, which correlates with the higher electron donor ability of the R groups, and consequently, with the higher donor ability of the alkyne. Furthermore, this higher donor ability of the alkyne with *EDGs* is responsible for

Table 5.1 Relative Gibbs energies in *DCM* (ΔG_{DCM}, kcal · mol^{-1}) at 298 K for the carbopalladation mechanism with the different 4-substituted phenylacetylenes (R = H, CF_3, OMe, NMe_2)

Species	Substituent R			
	H	CF_3	OMe	NMe_2
1 + Base + alkyne	0.0	0.0	0.0	0.0
TS1	23.0	24.8	22.9	20.6
2 + PH_3	13.5	15.9	12.7	9.9
C-TS1	27.7	28.3	27.5	26.7
C-1	−6.2	−7.1	−5.7	−5.4
C-TS2	0.2	−1.8	0.2	−0.1
C-2	−20.0	−21.5	−19.5	−19.8
C-TS3	20.4	19.4	21.3	21.2
3 + $[Pd(PH_3)_2]$ + H-Base$^+$I$^-$	−21.5	−21.6	−21.3	−21.4

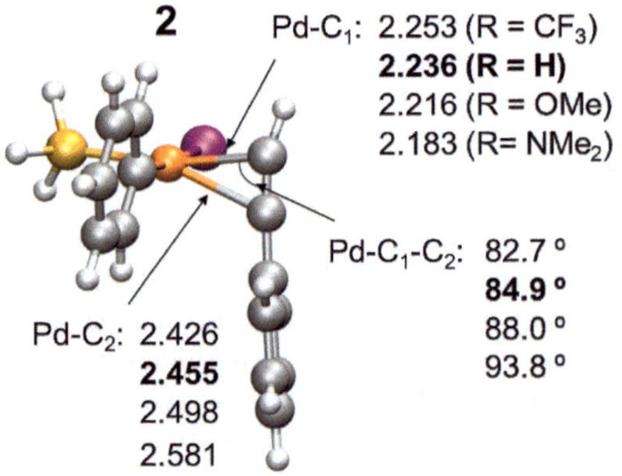

Fig. 5.7 Optimized structure for intermediate **2** with phenylacetylene (R = H). Distances (shown in Å) and angles in complex **2** with the other R groups are also shown

the higher stability of the corresponding complexes **2**, which feature shorter Pd–C_1 distances and higher Pd–C_1–C_2 angles with these substituents (Fig. 5.7).

Similarly to the case with R = H, the deprotonation of the alkene moiety in **C-2** with the other R groups has the highest energy barrier in the overall reaction pathway, with values ranging from 40.4 to 41.0 kcal · mol^{-1} (Table 5.1). Thus, for these substituted phenylacetylenes we could also conclude that the carbopalladation mechanism is also too energy-demanding to be operative under the reaction conditions, which agrees again with the experimental findings of Mårtensson et al. [43].

5.6 The Deprotonation Mechanism

With the carbopalladation mechanism ruled out as operative mechanism, the copper-free Sonogashira reaction through a deprotonation mechanism was next investigated. As commented in the introduction, for this mechanism two different alternatives have been proposed, namely the cationic and the anionic mechanisms (Fig. 5.3) [43]. This two mechanistic alternatives only differ in the order in which the steps in the deprotonation mechanism occur.

5.6.1 *The Cationic Mechanism*

The computed Gibbs energy profile for the copper-free Sonogashira reaction with phenylacetylene (R = H) through a cationic mechanism is shown in Fig. 5.8.

Unlike the carbopalladation mechanism, once complex **2** is formed, the substitution of the iodide by the phosphine ligand takes place with a relative energy barrier of 14.0 kcal · mol^{-1} (**DC-TS1**) giving rise to the ion-pair formed between the cationic Pd complex and the iodide (**DC-1**). Then, from this species the deprotonation of the alkyne by the external base occurs (**DC-TS2**, 25.9 kcal · mol^{-1}) yielding the intermediate **RE-1**, where the two organic groups are in a *cis* configuration. Finally,

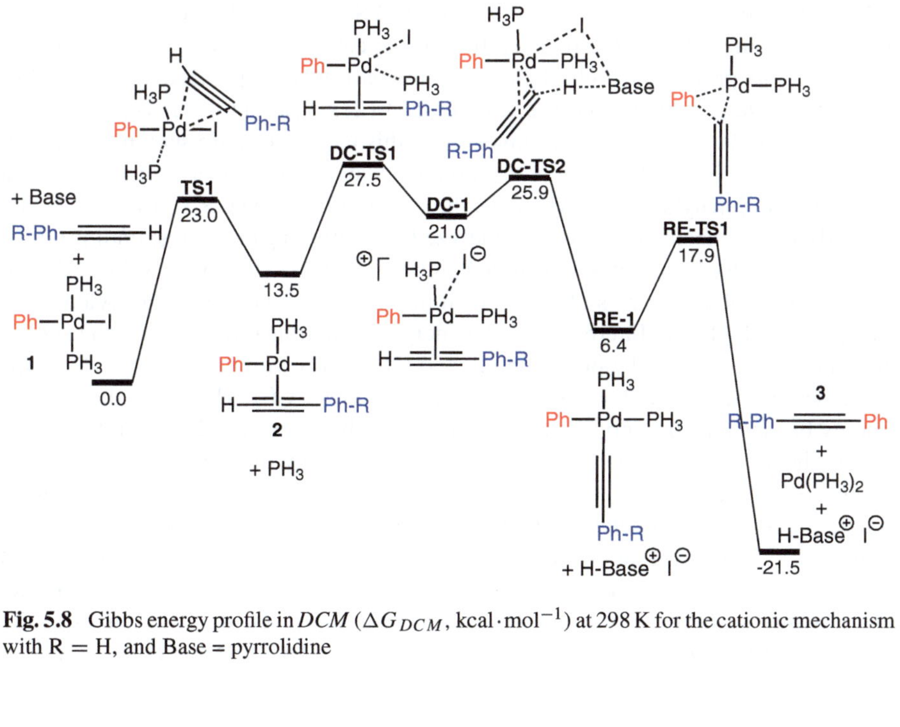

Fig. 5.8 Gibbs energy profile in *DCM* (ΔG_{DCM}, kcal·mol^{-1}) at 298 K for the cationic mechanism with R = H, and Base = pyrrolidine

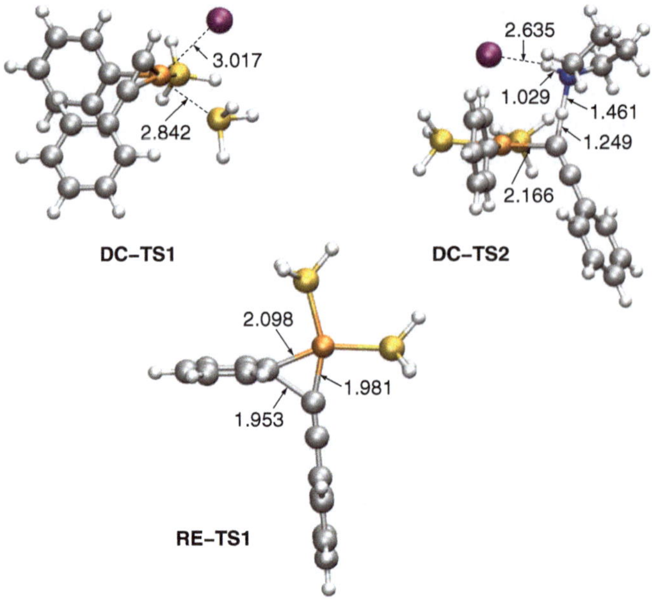

Fig. 5.9 Optimized structures for the transition states involved in the cationic mechanism with phenylacetylene (R = H). Distances are shown in Å

RE-1 undergoes reductive elimination via **RE-TS1** (17.9 kcal · mol^{-1}) resulting in the final product (**3**) and the regeneration of the catalytically active species. In the overall energy profile, the highest global energy barrier corresponds to the iodide-by-phosphine ligand substitution in **2** via **DC-TS1** (27.5 kcal · mol^{-1}). The optimized structures for the transition states involved in this mechanism are shown in Fig. 5.9.

Having established the cationic reaction pathway for the copper-free Sonogashira reaction with phenylacetylene (R = H), the effect of the alkyne R substituent on the overall cationic mechanism was next analyzed by computing the corresponding Gibbs energy profile for the other 4-substituted phenylacetylenes (R = CF_3, OMe, NMe_2). The relative Gibbs energies obtained with these 4-substituted phenylacetylenes are presented in Table 5.2. Interestingly, the energy of the transition state for the substitution of the iodide by the phosphine ligand (**DC-TS1**) with the different R groups increases in the order: R = NMe_2 < OMe < H < CF_3. However, the relative energy barrier for this step (the energy difference between **DC-TS1** and **2**) is practically the same for all the R groups (the highest energy difference between the R groups is 0.8 kcal · mol^{-1}), which indicates that the different stability of intermediates **2** is responsible for the differences in the global energy barriers. On the other hand, Table 5.2 shows that the deprotonation of the alkyne in **DC-1** through **DC-TS2** follows a similar trend that the previous iodide-for-phosphine substitution step (i.e. R = NMe_2 < OMe < CF_3 < H). In this case, however, the differences in the relative energy barriers (the energy difference between **DC-TS2** and **DC-1**) between

Table 5.2 Relative Gibbs energies in *DCM* (ΔG_{DCM}, kcal · mol^{-1}) at 298 K for the cationic mechanism with the different 4-substituted phenylacetylenes (R = H, CF_3, OMe, NMe_2)

Species	Substituent R			
	H	CF_3	OMe	NMe_2
1 + Base + alkyne	0.0	0.0	0.0	0.0
TS1	23.0	24.8	22.9	20.6
2 + PH_3	13.5	15.9	12.7	9.9
DC-TS1	27.5	29.1	26.4	23.7
DC-1	21.0	22.2	20.6	16.1
DC-TS2	25.9	25.5	25.3	21.5
RE-1	6.4	4.3	7.1	6.8
RE-TS1	17.9	15.5	18.8	18.8
3 + $[Pd(PH_3)_2]$ + H-Base$^+$I$^-$	−21.5	−21.6	−21.3	−21.4

the different R groups is slightly higher (the lowest energy difference between the R groups is 1.5 kcal · mol^{-1}). More specifically, the relative energy barriers for the deprotonation step increase in the following order: CF_3 < OMe ≈ H < NMe_2. This trend can be rationalized as follows: the presence of a *EWG* (i.e. R = CF_3) in the alkyne makes its proton more acidic and thus, the relative energy barrier for this step decreases. Conversely, the presence of an *EDG* (i.e. R = NMe_2) makes the proton of the alkyne less acidic, which increases the relative energy barrier. The cases with the model substituent (i.e. R = H) and the moderate *EDG* (i.e. R = OMe) have similar energy barriers and lie in between the other R groups.

Similarly to the reaction with phenylacetylene (R = H), the highest energy point in the overall energy profile corresponds to the substitution of the iodide by the phosphine ligand (**DC-TS1**) regardless of the R groups; therefore, this step has also the highest energy barrier for all the R groups. Furthermore, these energy barriers are lower for *EDGs* than for *EWGs*, indicating that the more *EDG* the faster the process via this cationic mechanism should be.

5.6.2 The Anionic Mechanism

As above stated, the reaction steps in the anionic mechanism take place in reverse order than in the cationic mechanism (Fig. 5.8). Thus, in the anionic mechanism the deprotonation of the alkyne by the external base in complex **2** occurs first, followed by the iodide-for-phosphine substitution. The Gibbs energy profile obtained for the copper-free Sonogashira reaction with phenylacetylene (R = H) through the anionic mechanism is shown in Fig. 5.10.

The deprotonation of the alkyne by the external base in complex **2** occurs through the transition state **DA-TS1** (23.0 kcal · mol^{-1}) and leads to the formation of the ion-pair (**DA-1**) formed between the anionic Pd complex and the protonated base.

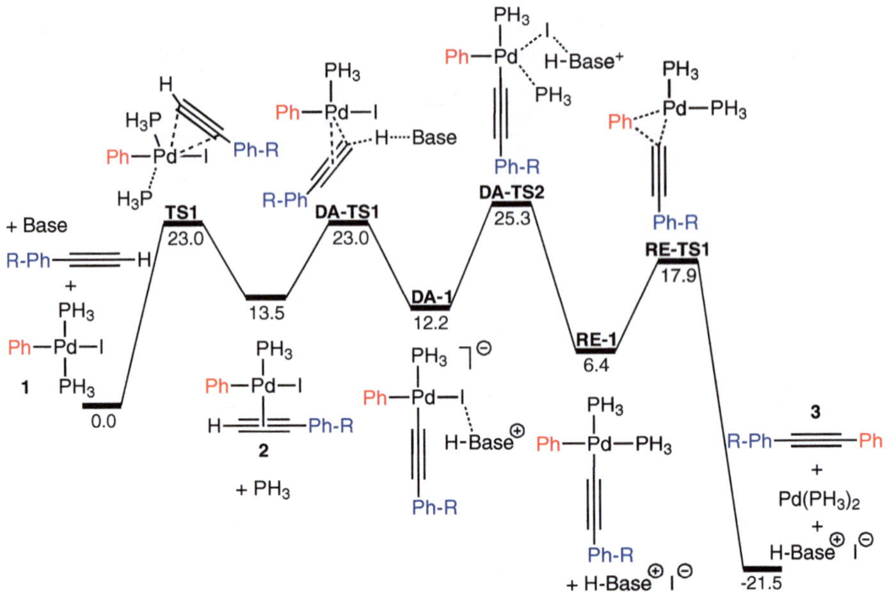

Fig. 5.10 Gibbs energy profile in *DCM* (ΔG_{DCM}, kcal·mol^{-1}) at 298 K for the anionic mechanism with R = H and Base = pyrrolidine

Subsequently, this species evolves to the next intermediate, **RE-1**, by a iodide-by-phosphine substitution via the transition state **DA-TS2** with a relative energy barrier of 13.1 kcal · mol^{-1}. Finally, **RE-1** undergoes reductive elimination giving rise to the coupled product and regenerating the catalytic species. The optimized structures for the transition states involved in this mechanism are shown in Fig. 5.11.

Overall, the substitution of the iodide by the phosphine ligand in **DA-1** through the transition state **DA-TS2** has a global energy barrier 2.3 kcal · mol^{-1} higher than

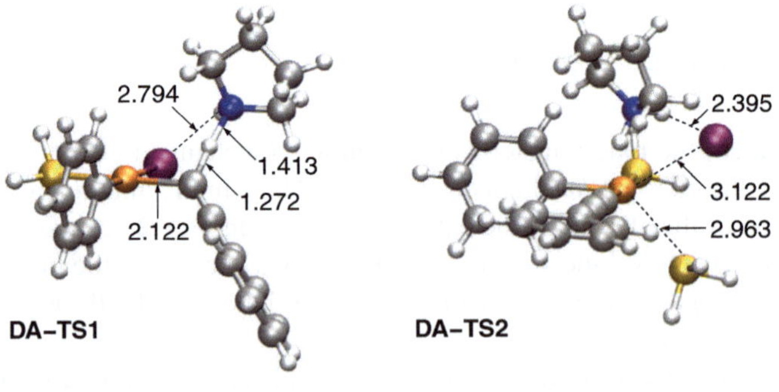

Fig. 5.11 Optimized structures for the transition states involved in the anionic mechanism with phenylacetylene (R = H). Distances are shown in Å

Table 5.3 Relative Gibbs energies in *DCM* (ΔG_{DCM}, kcal · mol^{-1}) at 298 K for the anionic mechanism with the different 4-substituted phenylacetylenes (R = H, CF_3, OMe, NMe_2)

Species	Substituent R			
	H	CF_3	OMe	NMe_2
1 + Base + alkyne	0.0	0.0	0.0	0.0
TS1	23.0	24.8	22.9	20.6
2 + PH_3	13.5	15.9	12.7	9.9
DA-TS1	23.0	23.1	22.3	19.3
DA-1	12.2	10.8	12.3	12.5
DA-TS2	25.3	24.6	25.8	25.6
RE-1	6.4	4.3	7.1	6.8
RE-TS1	17.9	15.5	18.8	18.8
3 + $[Pd(PH_3)_2]$ + H-$Base^{+}I^{-}$	−21.5	−21.6	−21.3	−21.4

the previous transition states (**TS1** and **DA-TS1**) and 7.4 kcal · mol^{-1} higher than the next transition state (**RE-TS1**). Therefore, the highest energy point in the anionic mechanism is **DA-TS2**. Interestingly, the energy of this transition state is a bit lower than the highest energy barrier in the cationic mechanism (**DC-TS1**, Fig. 5.8), which means that, for phenylacetylene (R = H), the anionic mechanism is favored.

With the anionic reaction pathway established for phenylacetylene (R = H), we next examined the effect of the alkyne R substituent on the overall anionic mechanism by computing the corresponding Gibbs energy profile for the other 4-substituted phenylacetylenes (R = CF_3, OMe, NMe_2). The relative Gibbs energies obtained with these 4-substituted phenylacetylenes are collected in Table 5.3.

Similarly to phenylacetylene (R = H), the highest energy barrier with the other 4-substituted phenylacetylenes corresponds to the substitution of the iodide by the phosphine ligand via **DA-TS2**. Moreover, the small energy differences between the different transition states **DA-TS2** (all of them within 1 kcal · mol^{-1}) suggest that the electronic nature of the alkyne R substituent does not have a significant effect on the reaction rates through this mechanism.

5.6.3 Cationic Mechanism Versus Anionic Mechanism

According to calculations, the highest energy barrier in both cationic and anionic mechanisms corresponds to the substitution of the iodide by the phosphine ligand. However, depending on the mechanism this substitution takes place before (i.e. in the cationic mechanism) or after (i.e. in the anionic mechanism) the deprotonation of the alkyne. In particular, in the case of the cationic mechanism this substitution occurs in complex **2** via **DC-TS1** (Fig. 5.8), whereas in the anionic mechanism it takes place in **DA-1** via **DA-TS2** (Fig. 5.10). The computed global Gibbs energy barriers for these processes with all the 4-substituted phenylacetylenes (Table 5.4) indicate that

Table 5.4 Global Gibbs energy barriers in *DCM* (ΔG_{DCM}, $kcal \cdot mol^{-1}$) at 298 K for the cationic and anionic mechanisms with the different 4-substituted phenylacetylenes (R = H, CF_3, OMe, NMe_2)

Substituent R	Global Gibbs energy barriers[a]	
	Cationic mechanism	Anionic mechanism
CF_3	29.1	**24.6**
H	27.5	**25.3**
OMe	26.4	**25.8**
NMe_2	**23.7**	25.6

[a]The Gibbs energy barriers for the lowest-energy deprotonation pathways with the different R groups are shown in bold

both mechanisms are feasible. Importantly, these values also suggest that depending on the electronic nature of the R substituent there might be a change in the reaction mechanism. More specifically, the highly electron withdrawing group R = CF_3 and the model substituent R = H, favor the anionic mechanism compared to the cationic mechanism by 4.5 and 2.2 $kcal \cdot mol^{-1}$, respectively. On the other hand, with the moderate electron donating group R = OMe this energy difference becomes smaller (0.6 $kcal \cdot mol^{-1}$) but still in favor of the anionic mechanism. Finally, with the highly electron donating group R = NMe_2 this energy difference is reversed favoring the cationic mechanism by 1.9 $kcal \cdot mol^{-1}$.

This predicted change in the reaction mechanism can be rationalized as follows: the first step in the cationic mechanism yields intermediate **2** (Fig. 5.8). As previously stated, in the case of *EWGs* (i.e. R = CF_3, H) or moderate *EDGs* (i.e. R = OMe) this species is less stable than with highly *EDGs* (i.e. R = NMe_2) (the energy difference range from 2.8 $kcal \cdot mol^{-1}$ to 6.0 $kcal \cdot mol^{-1}$). Thus, *EWGs* cause higher energy barriers because the following iodide-by-phosphine substitution step has a very similar relative energy barrier for all the R groups. In contrast, the first step in the anionic mechanism gives the intermediate **DA-1** (Fig. 5.10). In this case, the anionic charge on the Pd complex is better stabilized with *EWGs*, which offsets the energy gain in complex **2** with the highly *EDGs* and leads to lower energy barriers compared to the cationic mechanism. This results in a preference for the anionic mechanism with *EWGs* (i. e. R = CF_3, H) or moderate *EDGs* (i.e. R = OMe) and a preference for the cationic mechanism with highly *EDGs* (i.e. R = NMe_2).

5.7 An Alternative Mechanism: The Ionic Mechanism

While carrying out this mechanistic study on the copper-free Sonogashira reaction, we reported the combined experimental and theoretical study on the transmetalation step in the Negishi cross-coupling reaction [65] that has been described in Chap. 4. Therein, we demonstrated that an external coordinating ligand (i.e. $PMePh_2$, THF) can easily replace the chlorine from the complex *trans*-[$PdMeCl(PMePh_2)_2$] (com-

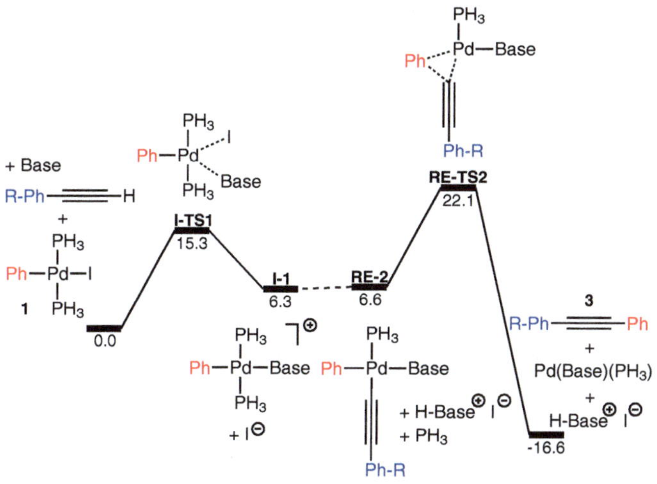

Fig. 5.12 Gibbs energy profile in *DCM* (ΔG_{DCM}, kcal · mol^{-1}) at 298 K for the ionic mechanism with R = H, and Base = pyrrolidine

plex analogous to **1**) through a mechanism that we dubbed as ionic mechanism (Figs. 4.17 and 4.19). Based on these results and based on the fact that the coordination of the alkyne in the copper-free Sonogashira reaction requires a considerably high energy barrier (higher than 20 kcal · mol^{-1}), we decided to compute a mechanism analogous to that ionic mechanism for the copper-free Sonogashira reaction. We labeled this mechanism as *ionic mechanism* owing to its similarities with the one reported for the Negishi coupling [65]. The computed Gibbs energy reaction profile via this mechanism with phenylacetylene (R = H) is shown in Fig. 5.12.

Unlike the cationic and anionic mechanisms, where the iodide is always expelled after the coordination of the alkyne, in the ionic mechanism it is replaced by the base in the first step. This process occurs through the transition state **I-TS1** and results in the formation of the cationic Pd complex **I-1**. Importantly, the energy barrier required for this process is 15.3 kcal·mol^{-1}, which is much lower (at least 5kcal · mol^{-1} lower) than the energy required for the coordination of the alkyne to **1** via **TS1** (Fig. 5.10). At this point, and based on the fact that the reaction is usually carried out with a large excess of base, we considered that the corresponding phenylacetylide may be present in solution as a result of the acid-base reaction. According to calculations, the phenylacetylide can replace one of the phosphine ligands in **I-1** through a barrier-less process.[1] This substitution without barrier yields the nearly isoenergetic species **RE-2**, which directly evolves to the desired alkyne (**3**) by common reductive elimination via **RE-TS2** (22.1kcal · mol^{-1}). The optimized structures of the transition states involved in this ionic mechanism are shown in Fig. 5.13.

[1] This was confirmed by optimizing the full system in dichloromethane starting with the phenylacetylene at 4 Å far from the cationic complex **I-1**.

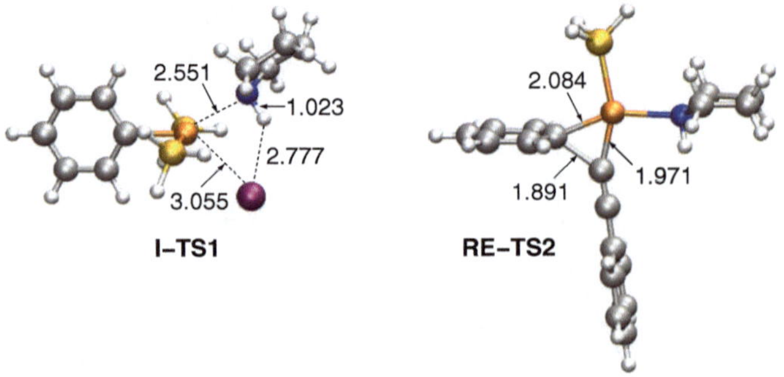

Fig. 5.13 Optimized structures for the transition states involved in the ionic mechanism with phenylacetylene (R = H). Distances are shown in Å

Overall, the reaction through this ionic mechanism is exergonic by 16.6 kcal · mol^{-1}. The step with the highest energy barrier corresponds to the reductive elimination step from complex **RE-2** via **RE-TS2** with an energy of 22.1 kcal·mol^{-1}. This energy barrier is lower than the ones found for the cationic and anionic mechanism, which suggests that this mechanism, in principle, should be favored. However, we have to take into account that this ionic mechanism depends on the concentration of the phenylacetylide present in solution, which is also directly linked to the concentration of the base. On this issue, the theoretical calculation of the *TOF* using the scheme developed by Kozuch et al. and by means of the *AUTOF* program [66–68] showed that the ratio between the anionic and the ionic mechanisms with phenylacetylene (R = H) is 2:1, thus indicating that both mechanisms are competitive.

As for the other analyzed mechanisms, after having established this ionic reaction pathway for phenylacetylene (R = H), we next examined the effect of the alkyne R substituent on the overall ionic mechanism. The relative Gibbs energies obtained in the computed Gibbs energy profiles with the other 4-substituted phenylacetylenes (R = CF_3, OMe, NMe_2) are presented in Table 5.5.

As shown in Table 5.5, for all the R substituents the highest energy barrier corresponds to the reductive elimination step through the transition state **RE-TS2**. The

Table 5.5 Relative Gibbs energies in *DCM* (ΔG_{DCM}, kcal · mol^{-1}) at 298 K for the ionic mechanism with the different 4-substituted phenylacetylenes (R = H, CF_3, OMe, NMe_2)

Species	Substituent R			
	H	CF_3	OMe	NMe_2
1 + Base + alkyne	0.0	0.0	0.0	0.0
I-TS1	15.3	15.3	15.3	15.3
I-1	6.3	6.3	6.3	6.3
RE-2	6.6	3.5	6.0	6.4
RE-TS2	22.1	18.5	22.6	21.8
3 + [Pd(Base)(PH_3)] + H-Base$^+$I$^-$	−16.6	−16.7	−16.4	−16.5

calculated energy barriers for this process with the different R groups rank from 18.5 (R = CF_3) to 22.6 kcal · mol^{-1} (R = OMe), which indicates that this ionic mechanism might be competitive with the cationic and anionic mechanisms. As above mentioned, the reaction rate through this ionic mechanism depends on the concentration of acetylide present in solution, which in turn depends on the concentration of the base. In other words, this mechanism depends on the base[2] and the acidity of the terminal alkyne proton, which at the same time depends on the electron withdrawing ability of the alkyne R substituent. Thus, the reaction through this mechanism is expected to be faster when alkynes bearing *EWGs* are used.

5.8 The Effect of the Alkyne R Substituent from Experiments

The theoretical calculations presented so far demonstrate that the carbopalladation mechanism is not operating under the reaction conditions. Furthermore, calculations also show that the other three investigated mechanisms (i.e. cationic, anionic and ionic mechanisms) may have competitive rates. Thus, a change on the reaction conditions (i.e. solvent, ligands, substrates, base, etc.) might favor one or another mechanism.

Regarding the effect of the alkyne R substituent on the analyzed mechanisms, theoretical results show relatively small energy differences in the highest Gibbs energy barriers with the different 4-substituted phenylacetylenes (Tables 5.4, 5.5). Hence, we decided, in collaboration with the group of Prof. Nájera, to carry out experimental copper-free Sonogashira reactions of 1-fluoro-4-iodobenzene with the 4-substituted phenylacetylenes used for the computational study for comparison. These couplings were performed in dichloromethane with $PdCl_2(PPh_3)_2$ (2 mol %) as catalyst, pyrrolidine (6 eq.) as base, at room temperature and under Ar atmosphere.[3] Then, for these reactions, the values of conversion (%) of the desired alkyne **3** versus time were obtained by monitoring the reactions by ^{19}F NMR spectroscopy (Fig. 5.14).

According to the conversion/time data plotted in Fig. 5.14 the reaction rate increases with the electron withdrawing ability of the R group (R = NMe_2 $<$ OMe $<$ H $\approx$ CF_3). Thus, the more acidic the terminal alkyne proton is, the higher the reaction rate is; this experimental trend is supported by the ionic mechanism. Notice, however, that if we compare the ratio of conversion of **3** after 1h for the fastest (R = CF_3 or H) and the slowest (R = NMe_2) reaction, it is about 4:1. This ratio in terms of activation energies corresponds to an energy difference lower than 1 kcal · mol^{-1}, which means that the activation energies for the copper-free Sonogashira reactions with the different 4-substituted phenylacetylenes are very similar, in agreement with theoretical calculations. In view of these results, there is not a clear preference for either the deprotonation or the ionic mechanism. Hence, the precise mechanism for a coupling reaction needs to be evaluated in detail for each partic-

[2] The 2:1 ratio between base and alkyne molecules in this ionic mechanism is compatible with the common reaction conditions, where an excess of base is frequently added.

[3] For further details on the experimental reaction conditions, see ref [46].

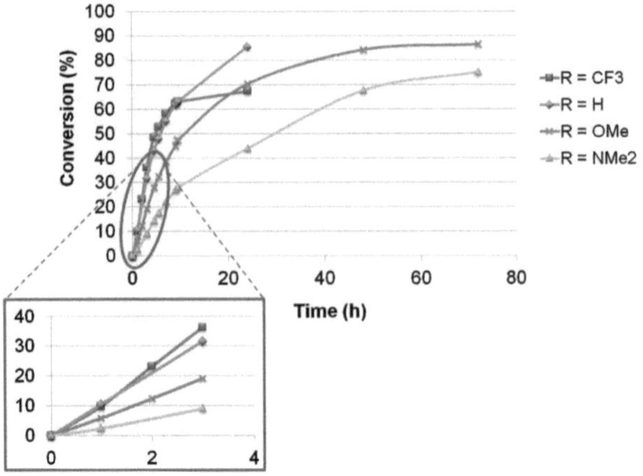

Fig. 5.14 Conversion/time data obtained by ^{19}F NMR for the Sonogashira reaction with the analyzed 4-substituted phenylacetylenes (R = H, CF_3, OMe, NMe_2)

ular case, and competitive mechanisms may take place together. The analysis on a model system for the Suzuki cross-coupling reaction gave rise to similar conclusions regarding the operative reaction mechanism [54].

5.9 Conclusions

In the study presented in this chapter, the reaction mechanism for the model copper-free Sonogashira reaction between iodobenzene and several 4-substituted phenylacetylenes (R = H, CF_3, OMe, NMe_2) was investigated by means of *DFT* calculations. Importantly, to the best of our knowledge, this study was the first theoretical study that investigated all the reported mechanistic proposals for the copper-free Sonogashira reaction.

The theoretical results on the carbopalladation mechanism showed that this pathway has a very high energy barrier, which indicated that this mechanism could not be operating under the reaction conditions. On the other hand, the calculated Gibbs energy barriers for the cationic and anionic alternatives proposed for the deprotonation mechanism, [43] revealed that both alternatives are feasible. Moreover, calculations predicted that one or the other reaction mechanism may be favored depending on the electronic nature of the alkyne R substituent. Thus, *EWGs* (R = CF_3, H) or moderate *EDGs* (R = OMe) may favor the anionic mechanism, whereas highly *EDGs* (R = NMe_2) may favor the cationic mechanism. These predicted differences can be attributed to the different stability of the intermediates that precede the highest energy barrier, which depends on the R substituent. These results are in agreement with the reported experimental work of Mårtensson et al. [43].

This study also demonstrated that the presence of an excess of a coordinating ligand like pyrrolidine, the base, and the presence of phenylacetylide opens a new reaction pathway for the copper-free Sonogashira reaction: the ionic mechanism. In this mechanism the role of the base is twofold: substituting the halide and helping the formation of the acetylide species. The theoretical results showed that this mechanism is competitive with the analyzed cationic and anionic mechanisms and that it may lead to higher reaction rates with terminal alkynes bearing *EWGs*, which agreed with our reported experiments.

In summary, this study revealed that in the copper-free Sonogashira reaction, like in other cross-coupling reactions (i.e. Stille, Negishi), there are several reaction pathways that may have competitive rates and a change on the reaction conditions (e.g. solvent, ligands, substrates, base) might favor one over the other ones. Thus, the general conclusion that we can draw from this work is that a detailed study of a particular copper-free Sonogashira reaction is required to assess which of the herein analyzed mechanisms is favored. This combined theoretical and experimental study on a model Pd-catalyzed copper-free Sonogashira reaction mapped out the reaction scenario and showed the complexity of this process.

References

1. Sonogashira, K.: J. Organomet. Chem. **653**, 46 (2002)
2. Negishi, E.-I., Anastasia, L.: Chem. Rev. **2003**, 103 (1979)
3. Tykwinski, R.R.: Angew. Chem. Int. Ed. **42**, 1566 (2003)
4. Chinchilla, R., Nãijera, C.: Chem. Rev. **107**, 874 (2007)
5. Chinchilla, R., Nãijera, C.: Chem. Soc. Rev. **40**, 5084 (2011)
6. Bunz, U.H.F.: Chem. Rev. **100**, 1605 (2000)
7. Nicolaou, K.C., Bulger, P.G., Sarlah, D.: Angew. Chem. Int. Ed. **44**, 4442 (2005)
8. Müller, T.J., D'Souza, D.M.: Pure Appl. Chem. **80**, 609 (2008)
9. Schiedel, M.S., Briehn, C.A., Bäuerle, P.: J. Organomet. Chem. **653**, 200 (2002)
10. Hortholary, C., Coudret, C.: J. Org. Chem. **68**, 2167 (2003)
11. Zapf, A., Beller, M.: Top. Catal. **19**, 101 (2002)
12. Torborg, C., Beller, M.: Adv. Synth. Catal. **351**, 3027 (2009)
13. Dieck, H.A., Heck, F.R.: J. Organomet. Chem. **93**, 259 (1975)
14. Cassar, L.: J. Organomet. Chem. **93**, 253 (1975)
15. Kotora, M., Takahashi, T.: In: Negishi, E.-I., de Meijere, A. (eds.) Handbook of Organopalladium Chemistry for Organic Synthesis, p. 973. Wiley-Interscience, New York (2002)
16. Bräse, S., de Meijere, A.: In: Negishi, E.-I., de Meijere, A. (eds.) Handbook of Organopalladium Chemistry for Organic Synthesis, p. 1133. Wiley-Interscience, New York (2002)
17. Sonogashira, K., Tohda, Y., Hagihara, N.: Tetrahedron Lett. **16**, 4467 (1975)
18. Ljungdahl, T., Pettersson, K., Albinsson, B., Mårtensson, J.: J. Org. Chem. **71**, 1677 (2006)
19. Gelman, D., Buchwald, S.L.: Angew. Chem. Int. Ed. **42**, 5993 (2003)
20. Siemsen, P., Livingston, R.C., Diederich, F.: Angew. Chem. Int. Ed. **39**, 2633 (2000)
21. Fukuyama, T., Shinmen, M., Nishitani, S., Sato, M., Ryu, I.: Org. Lett. **4**, 1691 (2002)
22. Soheili, A., Albaneze-Walker, J., Murry, J.A., Dormer, P.G., Hughes, D.L.: Org. Lett. **5**, 4191 (2003)
23. NÃijera, C., Gil-Moltó, J., Karlström, S., Falvello, L.R.: Org. Lett. **5**, 1451 (2003)
24. Rau, S., Lamm, K., Görls, H., Schöffel, J., Walther, D.: J. Organomet. Chem. **689**, 3582 (2004)
25. Li, J.-H., Zhang, X.-D., Xie, Y.-X.: Eur. J. Org. Chem. **2005**, 4256 (2005)

26. Gil-Moltó, J., Nãijera, C.: Eur. J. Org. Chem. **19**, 4073 (2005)
27. Yi, C., Hua, R.: Catal. Commun. **7**, 377 (2006)
28. de Lima, G.P., Antunes, O.A.C.: Tetrahedron Lett. **49**, 2506 (2008)
29. Komãiromi, A., Tolnai, G., Novãik, Z.: Tetrahedron Lett. **49**, 7294 (2008)
30. Alonso, D.A., Nãijera, C., Pacheco, M.C.: Tetrahedron Lett. **43**, 9365 (2002)
31. Cheng, J., Sun, Y.-H., Wang, F., Guo, M.-J., Xu, J.-H., Pan, Y., Zhang, Z.-G.: J. Org. Chem. **69**, 5428 (2004)
32. Ruiz, J., Cutillas, N., López, F., López, G., Bautista, D.: Organometallics **25**, 5768 (2006)
33. Komãiromi, A., Novãik, Z.: Chem. Commun. **40**, 4968 (2008)
34. John, A., Shaikh, M.M., Ghosh, P.: Dalton Trans. **47**, 10581 (2009)
35. Torborg, C., Huang, J., Schulz, T., Schäffner, B., Zapf, A., Spannenberg, A., Börner, A., Beller, M.: Chem. Eur. J. **15**, 1329 (2009)
36. Liang, B., Dai, M.-J., Chen, J.-H., Yang, Z.: J. Org. Chem. **70**, 391 (2005)
37. Li, J.-H., Liang, Y., Xie, Y.-X.: J. Org. Chem. **70**, 4393 (2005)
38. Carpita, A., Ribecai, A.: Tetrahedron Lett. **50**, 204 (2009)
39. Urgaonkar, S., Verkade, J.G.: J. Org. Chem. **69**, 5752 (2004)
40. Liang, Y., Xie, Y.-X., Li, J.-H.: J. Org. Chem. **71**, 379 (2006)
41. Amatore, C., Bensalem, S., Ghalem, S., Jutand, A., Medjour, Y.: Eur. J. Org. Chem. **2004**, 366 (2004)
42. Tougerti, A., Negri, S., Jutand, A.: Chem. Eur. J. **13**, 666 (2007)
43. Ljungdahl, T., Bennur, T., Dallas, A., Emtenas, H., Mårtensson, J.: Organometallics **27**, 2490 (2008)
44. Chen, L.-P., Hong, S.-G., Hou, H.-Q.: Chin. J. Struct. Chem. **27**, 1404 (2008)
45. Sikk, L., Tammiku-Taul, J., Burk, P.: Organometallics **30**, 5656 (2011)
46. Garcã-a-Melchor, M., Pacheco, M.C., Nãijera, C., Lledós, A., Ujaque, G.: ACS Catal. **2**, 135 (2012)
47. Becke, A.D.: J. Chem. Phys. **98**, 5648 (1993)
48. Lee, C., Yang, W., Parr, R.G.: Phys. Rev. B **37**, 785 (1988)
49. Frisch, M.J., Trucks, G.W., Schlegel, H.B., Scuseria, G.E., Robb, M.A., Cheeseman, J.R., Montgomery, J.A. Jr., Vreven, T., Kudin, K.N., Burant, J.C., Millam, J.M., Iyengar, S.S., Tomasi, J., Barone, V., Mennucci, B., Cossi, M., Scalmani, G., Rega, N., Petersson, G.A., Nakatsuji, H., Hada, M., Ehara, M., Toyota, K., Fukuda, R., Hasegawa, J., Ishida, M., Nakajima, T., Honda, Y., Kitao, O., Nakai, H., Klene, M., Li, X., Knox, J.E., Hratchian, H.P., Cross, J.B., Bakken, V., Adamo, C., Jaramillo, J., Gomperts, R., Stratmann, R.E., Yazyev, O., Austin, A.J., Cammi, R., Pomelli, C., Ochterski, J.W., Ayala, P.Y., Morokuma, K., Voth, G.A., Salvador, P., Dannenberg, J.J., Zakrzewski, V.G., Dapprich, S., Daniels, A.D., Strain, M.C., Farkas, O., Malick, D.K., Rabuck, A.D., Raghavachari, K., Foresman, J.B., Ortiz, J.V., Cui, Q., Baboul, A.G., Clifford, S., Cioslowski, J., Stefanov, B.B., Liu, G., Liashenko, A., Piskorz, P., Komaromi, I., Martin, R.L., Fox, D.J., Keith, T., Al-Laham, M.A. Peng, C.Y., Nanayakkara, A., Challacombe, M., Gill, P.M.W., Johnson, B., Chen, W., Wong, M.W. Gonzalez, C., Pople, J.A.: Gaussian 03, Revision E.01, Gaussian, Inc., Wallingford (2004)
50. Andrae, D., Häussermann, U., Dolg, M., Stoll, H., Preuss, H.: Theor. Chim. Acta. **77**, 123 (1990)
51. Ehlers, A.W., Bohme, M., Dapprich, S., Gobbi, A., Hollwarth, A., Jonas, V., Kohler, K.F., Stegmann, R., Veldkamp, A., Frenking, G.: Chem. Phys. Lett. **208**, 111 (1993)
52. Hollwarth, A., Bohme, M., Dapprich, S., Ehlers, A.W., Gobbi, A., Jonas, V., Kohler, K.F., Stegmann, R., Veldkamp, A., Frenking, G.: Chem. Phys. Lett. **208**, 237 (1993)
53. Check, C.E., Faust, T.O., Bailey, J.M., Wright, B.J., Gilbert, T.M., Sunderlin, L.S.: J. Phys. Chem. A **105**, 8111 (2001)
54. Braga, A.A.C., Ujaque, G., Maseras, F.: Organometallics **25**, 3647 (2006)
55. Nova, A., Ujaque, G., Maseras, F., Lledós, A., Espinet, P.: J. Am. Chem. Soc. **128**, 14571 (2006)
56. Álvarez, R., Pérez, M., Faza, O.N., de Lera, A.R.: Organometallics **27**, 3378 (2008)
57. Ariafard, A., Yates, B.F.: J. Am. Chem. Soc. **131**, 13981 (2009)

58. Pérez-RodrÃ-guez, M., Braga, A.A.C., GarcÃ-a-Melchor, M., Pérez-Temprano, M.H., Casares, J.A., Ujaque, G., de Lera, A.R., Álvarez, R., Maseras, F., Espinet, P.: J. Am. Chem. Soc. **131**, 3650 (2009)
59. Pérez-RodrÃ-guez, M., Braga, A.A.C., de Lera, A.R., Maseras, F., Álvarez, R., Espinet, P.: Organometallics **29**, 4983 (2010)
60. Thaler, T., Haag, B., Gavryushin, A., Schober, K., Hartmann, E., Gschwind, R.M., Zipse, H., Mayer, P., Knochel, P.: Nature Chem. **2**, 125 (2010)
61. Martin, R.L., Hay, P.J., Pratt, L.R.: J. Phys. Chem. A **102**, 3565 (1998)
62. Miertus, S., Scrocco, E., Tomasi, J.: J. Chem. Phys. **55**, 117 (1981)
63. Besora, M., Braga, A.A.C., Ujaque, G., Maseras, F., Lledós, A.: Theor. Chem. Acc. **128**, 639 (2011)
64. Casado, A.L., Espinet, P.: Organometallics **17**, 954 (1998)
65. GarcÃ-a-Melchor, M., Fuentes, B., Lledós, A., Casares, J.A., Ujaque, G., Espinet, P.: J. Am. Chem. Soc. **133**, 13519 (2011)
66. Kozuch, S., Shaik, S.: J. Am. Chem. Soc. **128**, 3355 (2006)
67. Kozuch, S., Shaik, S.: J. Phys. Chem. A **112**, 6032 (2008)
68. Uhe, A., Kozuch, S., Shaik, S.: J. Comput. Chem. **32**, 978 (2011)

Chapter 6
An Asymmetric Suzuki-Miyaura Reaction Mechanism

In order to succeed you must fail, so that you know what not to do the next time.

Anthony J. D'Angelo

6.1 Introduction

The Suzuki-Miyaura reaction, commonly known simply as Suzuki coupling, is one of the most practiced types of C–C cross-coupling reactions. This reaction is generally catalyzed by Pd complexes and consists in the carbon-carbon bond formation between an organic halide (or triflate), R^1–X, and an organoboron compound, R^2–BY_2, in the presence of a base (Scheme 6.1) [1–5]. The advantages of this reaction are numerous, such as the ready availability of reactants, the tolerance of a broad range of functional groups, the mild reaction conditions and high product yields, the low toxicity, etc. Consequently, the Suzuki-Miyaura coupling has been actively used not only in academic laboratories but also in industrial processes such as the large-scale synthesis of pharmaceuticals, natural products and fine chemicals [6–11].

As the rest of cross-coupling reactions, the Suzuki-Miyaura reaction mechanism consists in three general steps: oxidative addition, transmetalation, and reductive elimination. Unlike the former and the latter that are common to all cross-coupling reactions, the transmetalation is particular of this reaction as this is the step in which the organoboron compound actively participates. Thus, it is not surprising that since the discovery of the Suzuki-Miyaura coupling in 1979 [1], many efforts have been devoted to fully understand the mechanism of this singular step. For this process, two reaction pathways involving the required base have been proposed (Fig. 6.1): one consisting in the coordination of the base to the organoboron compound to form the organoboronate species, which undergoes nucleophilic attack on a palladium halide complex (pathway A, Fig. 6.1), and other consisting in the

M. G. Melchor, *A Theoretical Study of Pd-Catalyzed C–C Cross-Coupling Reactions*, Springer Theses, DOI: 10.1007/978-3-319-01490-6_6,

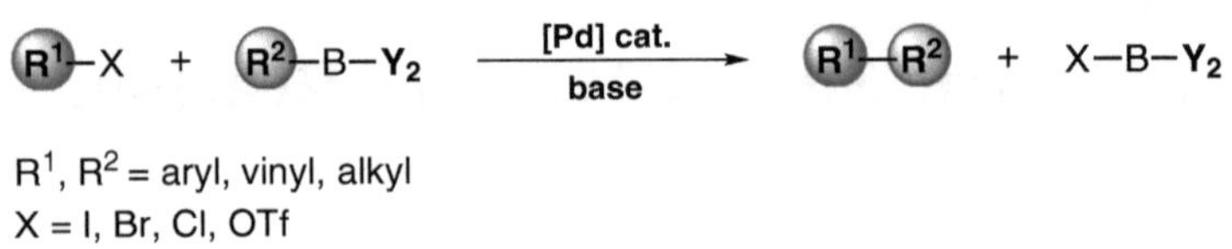

Scheme 6.1 General scheme for Pd-catalyzed Suzuki cross-coupling reaction

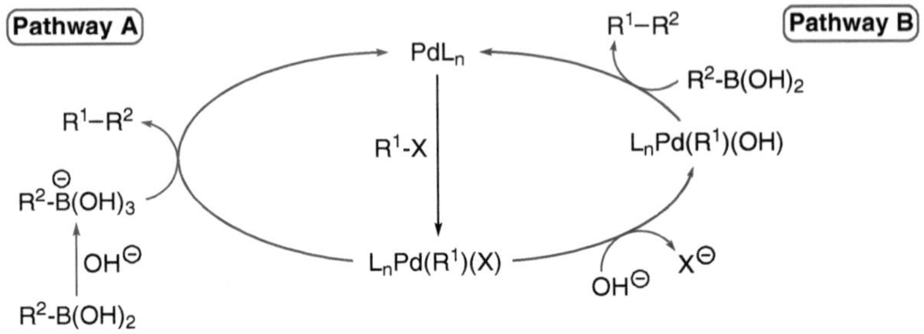

Fig. 6.1 Example for the proposed reaction pathways for the transmetalation step in the Suzuki-Miyaura cross-coupling reaction

base-for-halide substitution in the coordination sphere of the catalyst and the subsequent reaction with a neutral organoboron compound (Fig. 6.1, pathway B). Aimed at elucidating which of these two pathways is favored, both mechanisms were theoretically investigated by our group for a model Suzuki-Miyaura reaction using *trans*-[Pd(CH_2=CH)(Br)(PH_3)$_2$], CH_2=CH–B$(OH)_2$ and OH^- species as reactants [12]. The results derived from this study indicated that the pathway involving the organoboronate species (i.e. pathway A) is the most plausible one.

Thereafter, the transmetalation reaction has been generally taken for granted to go through the pathway A and, consequently, considered in further mechanistic studies on related Suzuki-Miyaura reactions [13–17]. The transmetalation via this pathway takes place in three steps (Fig. 6.2): first, the organoboronate species (generated by the reaction of the boronic acid with the base) replaces the halide in the coordination sphere of the Pd complex **S-1** (formed after the oxidative addition step) through the transition state **S-TS1**; subsequently, the resulting intermediate **S-3** undergoes intramolecular substitution of the OH ligand by the vinyl moiety via **S-TS2**, giving rise to the last intermediate **S-4**; and finally, this species evolves to the final products *trans*-[Pd(CH_2=CH)$_2$(PH_3)$_2$] (**S-5**) and B$(OH)_3$ by the concerted cleavage of the B–C bond and the coordination of the vinyl group to Pd through a single carbon atom (**S-TS3**).[1]

[1] The theoretical results obtained with this model system remained qualitatively valid when the PH_3 ligands and the vinyl groups were replaced by the commonly used PPh_3 ligands and phenyl groups [18], respectively. Importantly, in the transmetalation with phenyl groups, the reaction from **S-3** was found to take place in one transition state instead of in two (i.e. **S-TS2** and **S-TS3**).

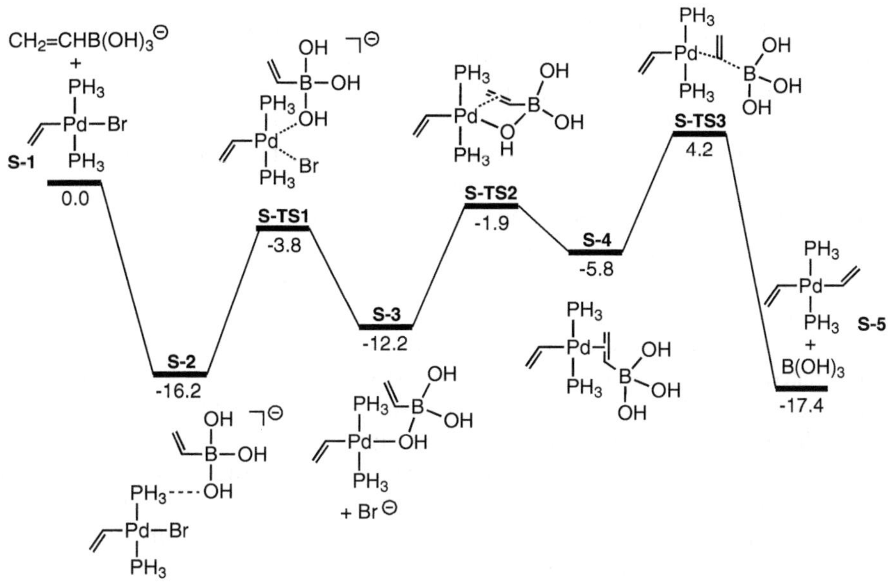

Fig. 6.2 Energy profile for the transmetalation of *trans*-[Pd(CH_2=CH)(Br)(PH_3)$_2$] (**S-1**) with CH_2=CH–B$(OH)_3^-$ via the pathway A [12]

Very recently, however, the proposal of the pathway A as the most plausible mechanism for transmetalation has been questioned with the appearance of a kinetic study by Hartwig et al. [19]. In particular, the results reported therein demonstrated that under typical Suzuki-Miyaura reaction conditions of weak base and aqueous solvent mixtures, the transmetalation process involves a palladium hydroxo complex and a neutral organoboron compound (pathway B, Fig. 6.1), rather than a palladium halide complex and the organoboronate species (pathway A, Fig. 6.1). The authors came to this conclusion based on the following experimental observations: *(i)* the populations of palladium halide and palladium hydroxo complex were found to be similar to each other; *(ii)* the populations of organoboron reagent and the corresponding organoboronate species resulted to be similar to each other in the presence of water and bases of the strength of carbonates; and *(iii)* the calculated rate for the reaction of the hydroxo complex with the neutral organoboron compound was several orders of magnitude faster than that for the reaction of halide complexes with the organoboronate species.

It should be noted, however, that the transmetalation reaction via the two proposed mechanisms (i.e. pathways A and B) leads to the formation of the same reaction intermediate (**S-3**, Fig. 6.2). Thus, assuming that this species precedes the step with the highest energy barrier (**S-TS3**, Fig. 6.2), it follows that the mechanism through which **S-3** is formed can be considered, to some extent, to be irrelevant.[2]

Similarly to other cross-coupling reactions [20–22], the steric and electronic effects induced by the ancillary ligands bound to the Pd catalyst on the Suzuki-Miyaura reaction has been also shown to be very important [23]. This interest mainly emerged from the improved efficiency and selectivity observed in Suzuki-Miyaura reactions with Pd catalysts possessing bulky and electron-rich phosphine ligands [24]. As commented in Chap. 1, the enhanced reactivity with this type of ligands is generally attributed to a combination of their electronic and steric properties that favors the stabilization of mono-ligated [PdL] intermediates, which are believed to actively participate in the catalytic cycle [25, 26]. Consequently, the effect of the ligands coordinated to the Pd catalyst on the Suzuki-Miyaura reaction has been extensively investigated in the last years by several groups, including ours [13, 27–31].

On the other hand, ligands bound to Pd catalysts are also of considerable importance because they can promote asymmetric catalytic transformations. In fact, one of the recent and growing interests in the Suzuki-Miyaura reaction has been the development of asymmetric versions for this reaction. In this context, the synthesis of axially chiral biaryls has attracted significant attention principally owing to their profound impact in organocatalysis [32, 33] and their biological and technological applications [34–36]. However, it should be mentioned that despite the considerable advances that have been made since the pioneering works of Buchwald [37] and Cammidge [38], the number of studies concerning asymmetric Suzuki-Miyaura cross-coupling to biaryls is still rather limited [39–45], specially those including theoretical calculations, which are very rare [46]. Hence, on view of this, we decided to investigate computationally the asymmetric Suzuki-Miyaura reaction recently reported by Fernández and Lassaletta [47], with the aim of providing a plausible explanation for the high yields and enantioselectivities observed therein. In particular, we focused our study on one of their reported examples for which an exceptional enantiomeric excess (% ee) and yield was obtained (see below).

In the following sections, a brief summary of the experimental work of Fernández and Lassaletta [47] will be presented, followed by our theoretical results obtained on this study. The reaction mechanism is divided in three steps: *(i)* oxidative addition, *(ii)* transmetalation, and *(iii)* reductive elimination, each of which presented in a separated section. It has to be mentioned, however, that this study is in process of completion. Therefore, the theoretical results and conclusions presented in this chapter are the ones obtained to date.

[2] As we will see later, this was assumed in the asymmetric Suzuki-Miyaura reaction investigated in this thesis.

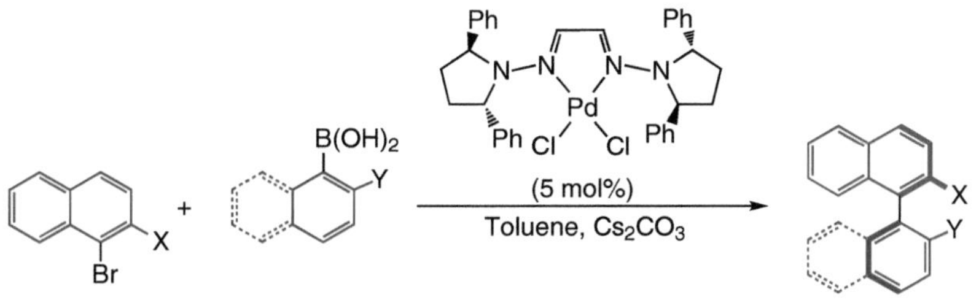

Scheme 6.2 Asymmetric Suzuki-Miyaura cross-coupling reactions

6.2 Experimental Data

In 2004, the experimental group of professors Fernández and Lassaletta reported an asymmetric Diels-Alder reaction catalyzed by a [Cu(OTf)$_2$/(bis-hydrazone)] catalyst in which the C$_2$-symmetric dialkylamino groups, making rotations around N–N bonds inconsequential, were deemed responsible for the high enantioselectivities obtained [48]. Based on these results and experimental [49, 50] and theoretical [51] precedents reported for the use of diimine and bis-hydrazone ligands in Suzuki-Miyaura reactions, the authors decided to explore the use of bis-hydrazone ligands in the development of a new asymmetric version for this reaction. In 2008, after preliminary experiments testing different chiral bis-hydrazones ligands in catalysts of the type [PdCl$_2$(bis-hydrazone)], the authors reported that the use of (*S,S*)-2,5-diphenylpyrrolidine-derived glyoxal bis-hydrazone ligand in conjunction with Cs_2CO_3 as base and toluene as solvent, allows the asymmetric Suzuki-Miyaura coupling of a broad variety of substrates in high yields and enantioselectivities (Scheme 6.2) [47].

Additionally, in an attempt to rationalize the high enantioselectivities and the absolute configuration observed for all the reactions, the authors proposed four possible reaction pathways based on rational orientations that ligands might adopt in the species formed in the course of the reaction (Fig. 6.3). Thus, in the oxidative addition step, two plausible species were considered to be generated: one with the X substituent of the organic halide pointing inside the page, and other with this group pointing outside (i.e. **I$_1$(A)** and **I$_1$(B)**, respectively). Subsequently, each of these two species may undergo transmetalation resulting in two new intermediates: one with the Y substituent of the organoboron reagent pointing inside the page, and other with this group pointing outside (i.e. **I$_2$(A)**-syn and **I$_2$(A)-anti** in the case of route A, and **I$_2$(B)-anti** and **I$_2$(B)-syn** in the case of route B). Finally, each of these species would afford the coupled product with the corresponding (*R*)- or (*S*)-configuration by common reductive elimination. Among these four pathways, the authors proposed the one going through **I$_2$(A)-anti** as the most favored one on the basis of the presumed presence of a higher number of π-stacking interactions between ligands. Furthermore, according to the authors the product derived from **I$_2$(A)-anti** would be generated by reductive elimination with counterclockwise conrotatory Pd–C(aryl)

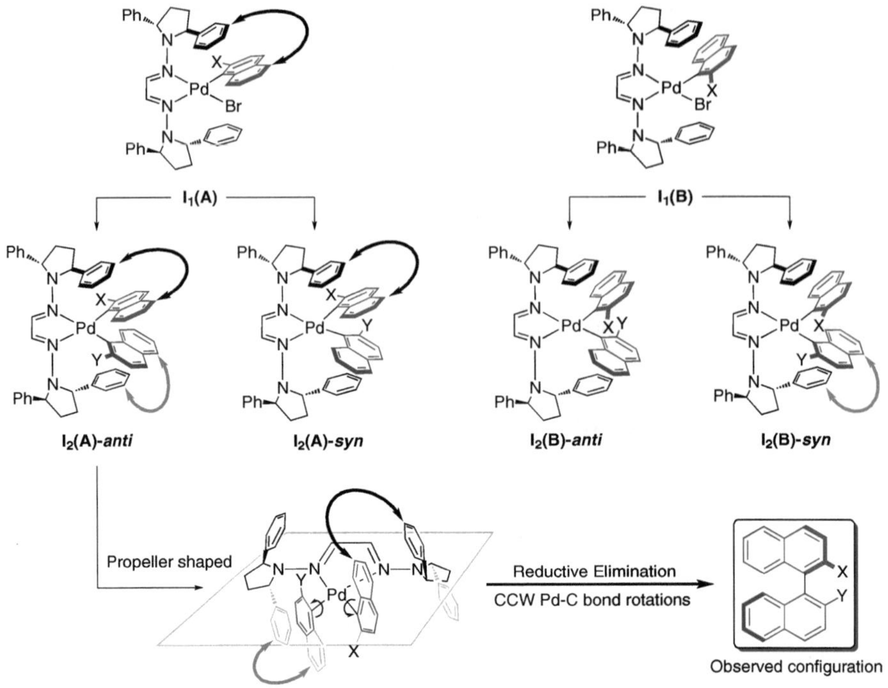

Fig. 6.3 Oxidative addition and transmetalation intermediates proposed to account for the configuration observed in the final product. Curved arrows indicate stabilizing π-stacking interactions. The route proposed as the most plausible one is depicted with bold arrows

bond rotations (favored against the opposite clockwise rotations for steric reasons), thus accounting for the observed configuration in the final product (Fig. 6.3).

Hence, prompted by these striking results, we decided to investigate theoretically the reported Suzuki-Miyaura coupling of 1-bromo-2-methylnaphthalene (Scheme 6.2, X = Me) with 1-naphthaleneboronic acid (Scheme 6.2, Y = H) [47]. We selected this coupling in purpose as it is one of the cases with the highest yield and enantiomeric excess reported therein, and furthermore, is one of the less computationally demanding. In particular, when this coupling was carried out at room temperature, a 80 % yield of the coupling (*R*)-product was obtained in 95 % ee (Fig. 6.4).[3]

As previously stated, the major objective of our ongoing theoretical study is to give an explanation of the origin of the enantioselectivities observed in the experiments. With this aim, the full catalytic cycle for the above mentioned coupling has been computed and the derived results analyzed. The computational details of this study and the results obtained so far, are presented in the next sections.

[3] The same coupling was also carried out at 80 °C affording the (*R*)-product at a higher yield (i.e. 98 %) but with a lower ee (i.e. 90 % (*R*)).

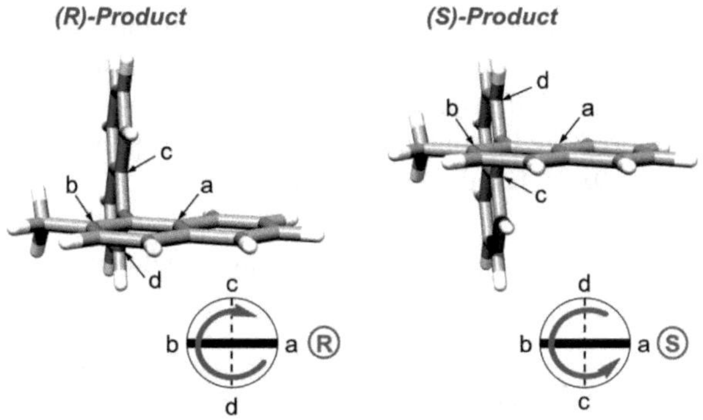

Fig. 6.4 Application of the method of designation [52] of the absolute (*R*),(*S*)-configuration to the investigated coupling product

6.3 Computational Details

All the calculations presented in this study have been performed at the *DFT* level of theory by means of the dispersion-corrected *M06* functional [53] and using the *Gaussian09* program [54]. We chose this functional for calculations based on the satisfactory results that we obtained in the mechanistic study on the transmetalation step in the Negishi cross-coupling reaction [55, 56], and the good description of non-covalent interactions that this functional has shown to provide [53]. The latter, is particularly important in this study because this type of interactions (e.g. π-stacking) are very likely to be present in reactions involving biaryls. Actually, the experimental study [47] pointed out to these interactions as responsible for the high enantioselectivities and the absolute configuration observed.

Importantly, this study has been carried without modeling the catalyst and the reagents employed for the experiments (Fig. 6.2, X = Me, Y = H). The B, C, and H atoms have been described using the standard 6-31G(d,p) basis set, while for the more electronegative O and N atoms the 6-31+G(d) basis set including diffuse functions has been employed. On the other hand, for Pd and Br atoms the *SDD* effective core potential [57] and its associated double-ζ basis set have been adopted. Additionally, for Pd f-polarization functions have been also added (exponent = 1.472) [58], whereas for Br both d-polarization (exponent = 0.428) and p-diffuse (exponent = 0.0376) functions [59] have been added.

All the geometries of the reactants, products, and transition states shown in this study have been fully optimized without symmetry constraints. These structures have been further characterized as minima or saddle points by computing the corresponding harmonic force constants and vibrational frequencies at the optimized geometries. In particular, the former species have been confirmed by having a full set of real frequencies, and the latter by presenting a unique imaginary frequency and correlating the corresponding reactants and products.

Solvent effects (i.e. toluene, $\varepsilon = 2.3741$) have been introduced by means of a continuum model, the *SMD* [60] solvation model[4] implemented in *Gaussian09*, performing single point calculations at all the optimized gas phase geometries.

All the energies presented throughout this study correspond to Gibbs energies in toluene (ΔG_{solv}) at 298.15 K calculated as follows:

$$\Delta G_{solv} = \Delta E_{solv} + (\Delta G_{gas} - \Delta E_{gas})$$

6.4 The Oxidative Addition Step

The oxidative addition is the first step in the catalytic cycle. In the particular coupling under study, this reaction takes place between the complex [Pd(bis-hydrazone)] (**1**) and 1-bromo-2-methylnaphthalene (**2**) (Scheme 6.2, X = Me). As proposed in the experiments [47], two different products can be envisaged in this process: one with the Me group of the organo halide **2** pointing inside the plane defined by palladium and their bound atoms, and other with that group pointing outside this plane (**I_1(A)** and **I_1(B)** in Fig. 6.3, respectively). Therefore, it follows that there are two possible approaches of **2** to the starting Pd complex **1** (approaches A and B giving rise to the calculated intermediates **OAA-I2** and **OAB-I2**, Fig. 6.5). In view of this, and in order to further locate the corresponding transition states for these two processes, we started by calculating the intermediates prior to these reactions. Interestingly, for both approaches, two different η^2-coordination modes of **2** to the Pd complex **1** were found, namely η^2-*a* and η^2-*j* (corresponding to **OAA-I1/OAB-I1** and **OAA-I1′/OAB-I1′**, respectively, in Fig. 6.5).[5] The optimized structures for the reaction intermediates with these η^2-coordination modes show shorter Pd–C distances (with the carbon atoms involved in this coordination) for the η^2-*a* coordination regardless of the approach; this can be attributed to the electron-donating effect of the Me group and the less steric hindrance that **2** exerts to the ligands in **1** adopting this coordination. Conversely, the distances in the computed intermediates for the two approaches presenting the same η^2-coordination have almost identical distances (with differences within 0.07 Å).

Subsequently, from the above optimized reaction intermediates, the corresponding transitions states for the oxidative addition reaction were computed (Fig. 6.6).[6] As expected, the located transitions states display a three-centered member ring formed between Pd, C_1 and Br atoms, and present an imaginary frequency that corresponds

[4] For further details on this model, see the last section of Chap. 2.

[5] The nomenclature adopted for these η^2-coordination modes is the one according to the *IUPAC* nomenclature for polycyclic aromatic hydrocarbons. For simplicity, we denoted the η^2-*j* coordination in the different intermediates shown in Fig. 6.5 by adding a prime symbol (′) to their names.

[6] The transition state starting from **OAA-I1** and **OAB-I1′** intermediates could not be located. Additional calculations are currently underway.

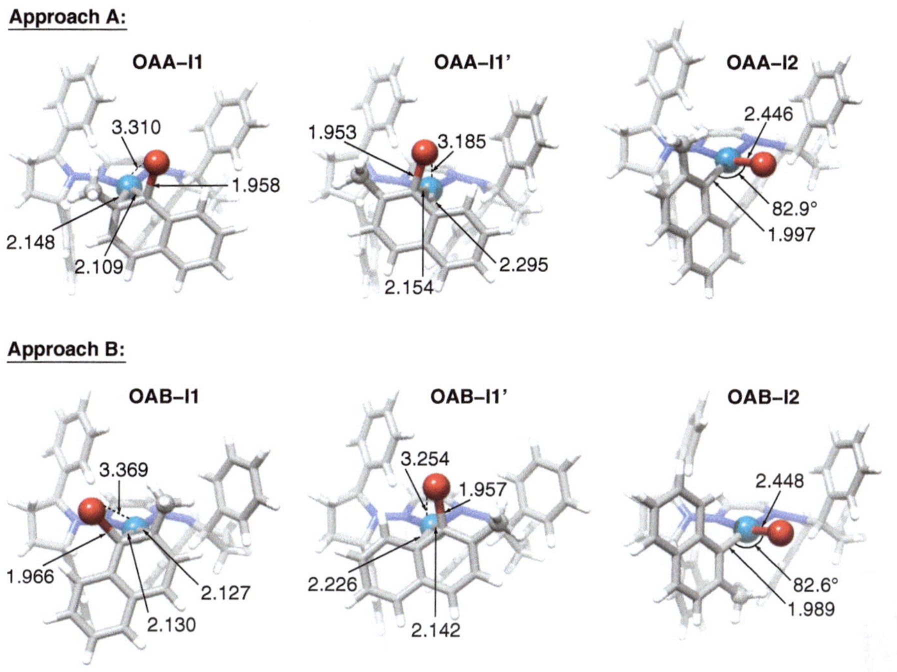

Fig. 6.5 Optimized structures for the intermediates and products involved in the oxidative addition reactions through the two possible approaches. Only the Br atom and the Me group of **2** are displayed using a ball-and-stick representation for a better distinction between these two approaches. Relevant distances are shown in Å

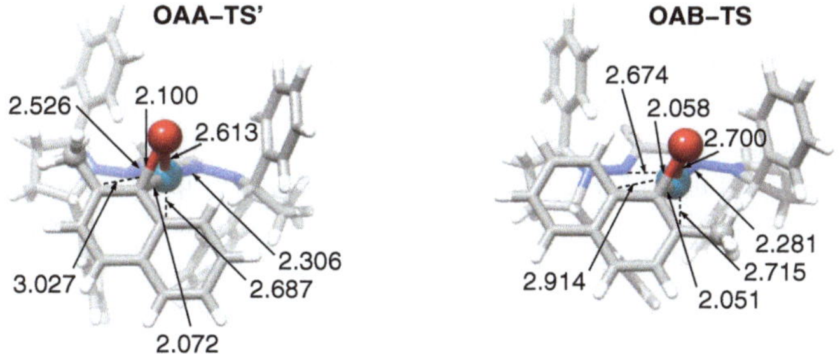

Fig. 6.6 Optimized structures for the transition states involved in oxidative addition reactions through the two possible approaches. Only the Br atom and the Me group of **2** are displayed using a ball-and-stick representation for a better distinction between these two approaches. Relevant distances are shown in Å

to the concerted cleavage of the C_1–Br bond and the coordination of **2** to **1** through only the C_1 atom.

The computation of the Gibbs energies in toluene for all the species involved in the oxidative addition step led to the Gibbs energy profile depicted in Fig. 6.7.

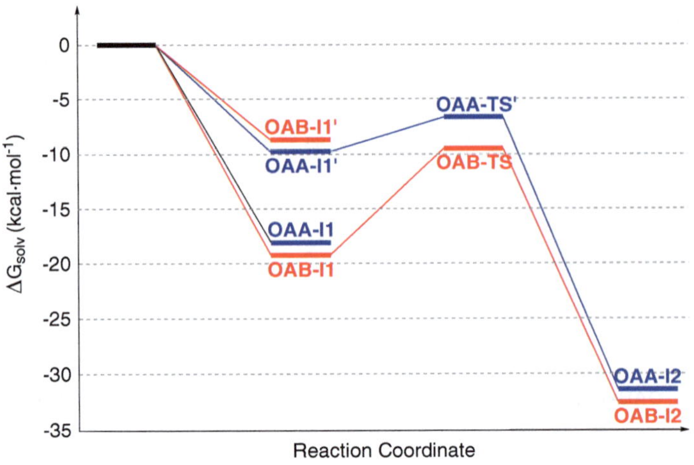

Fig. 6.7 Gibbs energy profile in toluene (ΔG_{solv}, kcal $\cdot$ mol^{-1}) at 298 K for the two possible oxidative addition reactions of **2** to the palladium complex **1**. Approach A is shown in *blue*, whereas approach B is shown in *red*

As far as the thermodynamics is concerned, the Gibbs energy profile show that both oxidative addition reactions through the two possible approaches are highly exergonic, with the product from the approach B (**OAB-I2**) 1.6 kcal $\cdot$ mol^{-1} more stable than the one from A (**OAA-I2**). Furthermore, the calculated Gibbs energies for the intermediates previous to the oxidative addition transition states indicate that there is a clear preference for the η^2-*a* coordination mode (**OAA-I1** and **OAB-I1**). This can be explained by the stronger Pd–C interactions (with the carbon atoms involved in this coordination) due to the electron-donating effect of the Me group and the lower steric hindrance between **2** and the ligands bound to Pd in **1** with this coordination. On the other hand, from the kinetic point of view, the Gibbs energy profile shows that the transition state **OAB-TS** has an energy value comparable to intermediates **OAA-I1**$'$ and **OAB-I1**$'$. Therefore, calculations suggest that the oxidative addition through the η^2-*a* coordination is the one taking place.

6.5 The Transmetalation Step

After the oxidative addition, the next step in the catalytic cycle is the transmetalation step. As we have seen in the introduction of this chapter, this step was recently investigated by our group [13] and, even though there is still some controversy over the role of the base, its reaction pathway is in general quite well understood. Thus, based on the generally accepted reaction mechanism (Fig. 6.2), we next investigated the transmetalation reaction between the two oxidative addition products (i.e. **OAA-I2** and **OAB-I2**) and 1-naphthaleneboronic acid (**3**) (Scheme 6.2, Y = H).

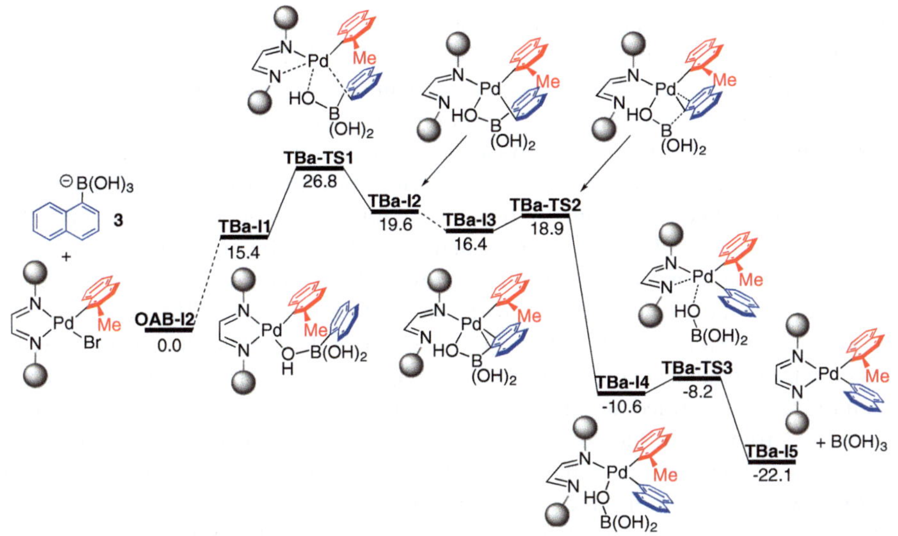

Fig. 6.8 Gibbs energy profile in toluene (ΔG_{solv}, kcal $\cdot$ mol^{-1}) at 298 K for the transmetalation reaction through the pathway *B-anti*. Substituents in the bis-hydrazone ligand have been simplified for clarity

According to the experimental work of Fernández and Lassaletta [47], two transmetalation products for each oxidative addition product can be expected on the basis of the two possible orientations of the organic group of the organoboronic species. These two products, in our case (Fig. 6.3, Y = H), correspond to the Pd complexes where the phenyl ring of naphthalene that is not coordinated to Pd (i.e. the "tail" of naphthalene) is pointing inside and outside the plain of the page. Consequently, there is a total of 4 possible routes for transmetalation. In order to keep the nomenclature for these routes consistent with the experimental work (Fig. 6.3), we named *"anti"* those routes where the two tails of the naphthalenes are pointing to opposite directions, and *"syn"* those routes where the two tails are pointing to the same direction. Thus, the name for the four transmetalation routes are: *A-anti*, *A-syn*, *B-anti*, and *B-syn*. In this chapter, since the reaction mechanism found for the four routes is basically the same, only the one for the *B-anti* route will be described herein.

The computation of the reaction pathway for the transmetalation through the route *B-anti* resulted in the Gibbs energy profile shown in Fig. 6.8.

The first intermediate in this profile is **TBa-I1**, which, as we previously explained, can be generated by either the substitution of the bromide by the organoboronate species (pathway A, Fig. 6.1), or by the OH-for-bromide substitution and subsequent coordination of the neutral organoboron species to the OH ligand (pathway B, Fig. 6.1). We also noted that, recently, some controversy has been stirred over which of these two processes is favored but that, in any case, they are not expected to contribute significantly to the observed reactivity as the step with the highest energy barrier corresponds to the cleavage of the B–C bond (**S-TS3**, Fig. 6.2). Thus, in this work we decided to do not further study this step and focus our attention on investigating the transmetalation reaction from **TBa-I1** (Fig. 6.8).

According to the generally accepted mechanism for aryls [18], once **TBa-I1** is formed, it undergoes transmetalation reaction trough a four-membered ring transition state in which the Pd–O and B–C bonds are broken at the same time that the Pd–C bond is formed. In our case, however, all the efforts devoted to locate and characterize such transition state were fruitless. Instead, we found the η^2-coordination of **3** to Pd and the substitution of one of the N ligands bound to Pd by one of the OH ligands of boron. The transition state for this process (**TBa-TS1**) requires a Gibbs energy barrier of 26.8 kcal $\cdot$ mol^{-1} and leads to the next intermediate **TBa-I2**, where **3** is coordinated to Pd in a η^2-*b* fashion (Fig. 6.9). Then, from this species, in order to undergo transmetalation the change from the η^2-*b* to the η^2-*j* coordination mode is required.[7] The intermediate resulting from this η^2-coordination change is **TBa-I3** (Fig. 6.9), from which the concerted cleavage of the B–C bond and the formation of the Pd–C bond occurs with a low Gibbs energy barrier (2.5 kcal $\cdot$ mol^{-1}) via **TBa-TS2**. This transition state gives rise to **TBa-I4**, which affords the final product of the transmetalation by the substitution of the borane by the N ligand through the transition state **TBa-TS3**. The Gibbs energy required for this last step is very low (1.6 kcal $\cdot$ mol^{-1}), yielding **TBa-I5** with the two organic groups in a *cis* disposition ready for the coupling by reductive elimination. The optimized structures for the most relevant species involved in the transmetalation step are shown in Fig. 6.9.

In summary, the calculated Gibbs energies depicted in Fig. 6.8 shows that the transmetalation reaction is exergonic by 22.1 kcal $\cdot$ mol^{-1}. The highest energy point within the overall Gibbs energy profile is **TBa-TS1** (26.8 kcal $\cdot$ mol^{-1}), which corresponds to the η^2-coordination of **3** to Pd and the concomitant substitution of one of the N ligands by one of the OH groups of boron (**TBa-TS1**, Fig. 6.9). As expected, the computed transition states analogous to **TBa-TS1** in the other three alternative transmetalation routes also provide the highest Gibbs energy barriers. These values, as well as the Gibbs energies for the transition states corresponding to **TBa-TS2**, are collected in Table 6.1.

As above mentioned, the step with the highest energy barrier in all the transmetalation routes is the one that involves the η^2-coordination of **3** to Pd and the substitution of one of the N ligands by one of the OH ligands of boron (**TS1**, Table 6.1). Consequently, the route *A-anti* that has the lowest energy for that step (23.9 kcal $\cdot$ mol^{-1}), is expected to be the most favorable route for transmetalation.

[7] This does not happen in the cases in which the tail of **3** is below the plain defined by Pd and their coordinated ligands, as the η^2-coordination that results from **TBa-TS1** is already the appropriate η^2-*b* coordination required in **TBa-TS2**. Hence, in the *A-anti* and *B-syn* routes no coordination change is needed and, accordingly, the transmetalation takes place directly from **TBa-I2** via the transition state **TBa-TS2**. In contrast, in the *A-syn* route the transmetalation occurs in the same way than in the *B-anti* route.

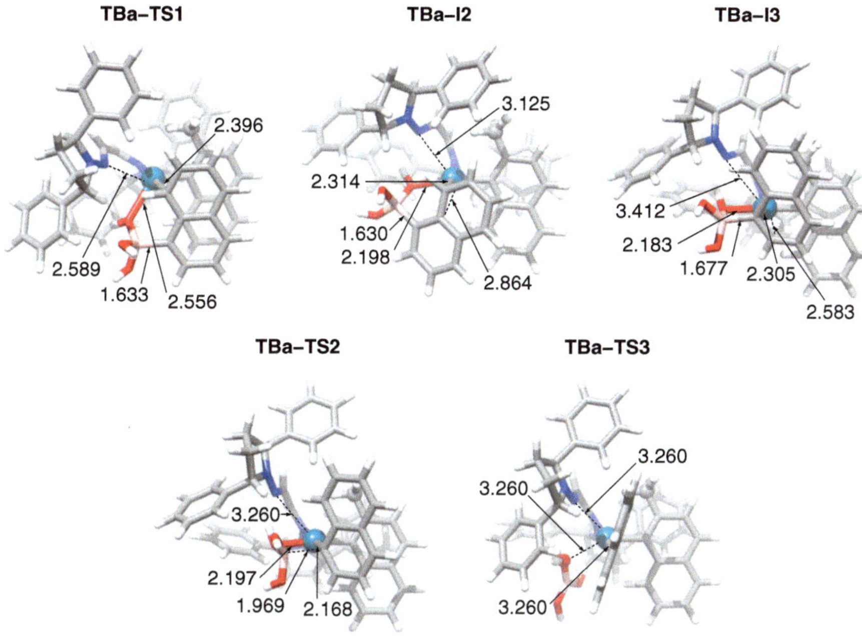

Fig. 6.9 Optimized structures for the transition states involved in transmetalation reaction through the route *B-anti*. Only the Me group of **2** and the oxygen atom coordinated to Pd in **TBa-TS3** are displayed using a ball-and-stick representation for clarity. Relevant distances are shown in Å

Table 6.1 Calculated Gibbs energy barriers in toluene (ΔG_{solv}, kcal $\cdot$ mol^{-1}) at 298 K for the transition states corresponding to **TBa-TS1** (i.e. **TS1**) and **TBa-TS2** (i.e. **TS2**) in the different transmetalation routes. The highest-energy transition states within these routes are shown in bold

	Route A[a]		**Route B**	
	A-anti	**A-syn**	**B-anti**	**B-syn**
TS1	**23.9**	**26.2**	**26.8**	**26.8**
TS2	21.6	20.7	18.9	22.6

[a] These Gibbs energy barriers are relative to the separated oxidative addition product **OAA-I2** and **3**, which are 1.6 kcal $\cdot$ mol^{-1} less stable than the separated oxidative addition product **OAB-I2** and **3**

6.6 The Reductive Elimination Step

The last step in the catalytic cycle that leads to the final coupling product with the defined stereochemistry and the concomitant regeneration of the catalytic species, is the reductive elimination. Thus, for the sake of completeness, the reductive elimination reactions starting from the products obtained in the four analyzed transmetalation routes (Fig. 6.10) were investigated. In particular, for each of these intermediates, the corresponding transition state and product involved in this process were computed. As expected, in all the calculated transition states for the different reductive eliminations, the formation of a three-membered ring between Pd and the two carbon

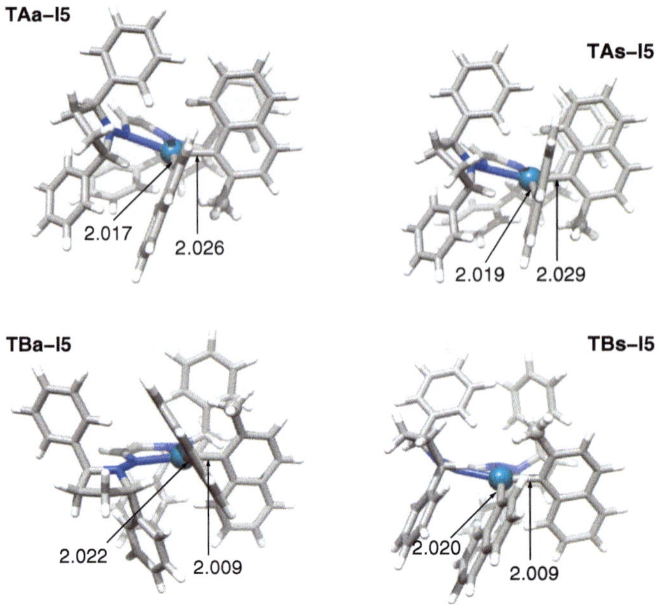

Fig. 6.10 Optimized structures for the different intermediates previous to the reductive elimination reactions. Only the Me group of **2** is represented in ball-and-stick for a better distinction between the different intermediates. The letters "A" and "B" in the labels of the intermediates refer to the routes A and B, respectively, whereas "a" and "s" refers to *-anti* or *-syn*, respectively. Relevant distances are shown in Å

atoms to be coupled, as well as an imaginary frequency according to this process, were found. The optimized transition state structure for the reductive elimination that provides the lowest Gibbs energy barrier (i.e route *A-syn*) is shown in Fig. 6.11.

According to calculations, all the analyzed reductive eliminations have reasonable Gibbs energy barriers, with values ranging from 14.6 (i.e. *A-syn*) to 17.4 $\text{kcal} \cdot \text{mol}^{-1}$ (i.e. *B-syn*) (Table 6.2).

As we stated in the beginning of this section, each of the transition states for the different routes leads to a coupled product with a defined stereochemistry. In particular, one of the routes (i.e. *A-anti*) results in the coupling product with an (*S*)-configuration, whereas the other three routes (i.e. *A-syn*, *B-anti* and *B-syn*) afford the (*R*)-coupling product (Table 6.2). This distribution of products is completely unexpected, since, a priori, two routes should give the (*R*)-product and two routes the (*S*)-product. Therefore, it follows that either the *B-anti* or the *B-syn* route should provide the (*S*)-product instead of the (*R*)-one. In principle, on the basis of the stereochemistry that results from the routes A (i.e. *A-anti* $\rightarrow$ *S*, and *A-syn* $\rightarrow$ *R*) one would expect the following stereochemistry for the routes B: *B-anti* $\rightarrow$ *R*, and *B-syn* $\rightarrow$ *S*. Nevertheless, for the *B-syn* route the (*R*)-product was obtained. The justification for this surprising result is given in the following section.

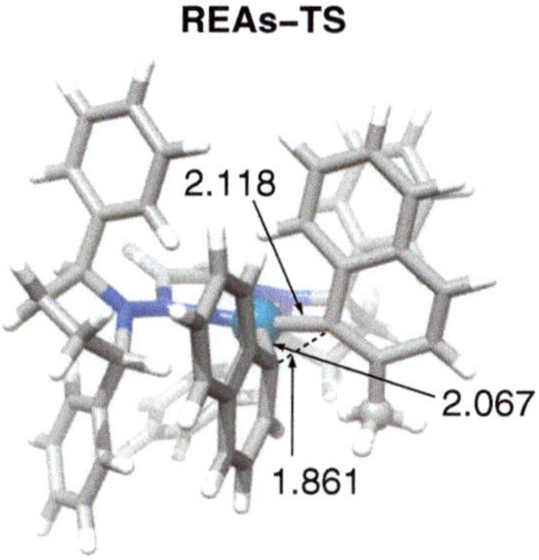

Fig. 6.11 Optimized transition state structure for the reductive elimination reaction through the route *A-syn*. Only the Me group of **2** is displayed using a ball-and-stick representation for clarity. Relevant distances are shown in Å

Table 6.2 Calculated Gibbs energy barriers in toluene (ΔG_{solv}, kcal $\cdot$mol^{-1}) at 298 K for the different reductive elimination reactions

Route A		**Route B**	
A-anti	**A-syn**	**B-anti**	**B-syn**
15.8 (*S*)	14.6 (*R*)	15.8 (*R*)	17.2 (*R*)

Gibbs energies are relative to the corresponding oxidative addition products. The stereochemistry of the product formed after reductive elimination is denoted in parenthesis

6.7 Where is the Enantioselectivity Defined?

In the previous section we have seen that the reductive elimination is the step that regenerates the catalytic species and affords the final product with a defined (*R*)- or (*S*)-configuration. However, it does not necessarily imply that the stereochemistry of the product is defined in that step. Actually, if one takes a look to the intermediates previous to reductive elimination (Fig. 6.10), one could (in principle) predict the absolute configuration of the final product. For instance, in the case of the intermediate of route *B-anti* (**TBa-I5**, Fig. 6.10) the tail of **3** is rather twisted toward the left, which ends up defining the (*R*)-configuration observed after reductive elimination (Table 6.2). In contrast, if the same ligand **3** were twisted toward the right, the C–C coupling from this intermediate would result in the (*S*)-product. Importantly, whether the tail of **3** will be right or left rotated will depend on its steric hindrance with the ligands coordinated to Pd. Consequently, the absolute (*R*)- or (*S*)-configuration of the coupling product will depend on steric effects (Fig. 6.12).

In the case of **TBa-I5**, Fig. 6.12 shows how the steric hindrance exerted by the Me group of **2** favors the rotation of the tail of **3** to the left, thus favoring the formation of the (*R*)-product. Similarly, in **TAa-I5**, the Me group of **2** also promotes the rotation of the tail of **3** to the left, and accordingly, the (*S*)-product is obtained. On the other hand, in **TAs-I5**, the Me group of **2** and the tail of **3** are oriented in

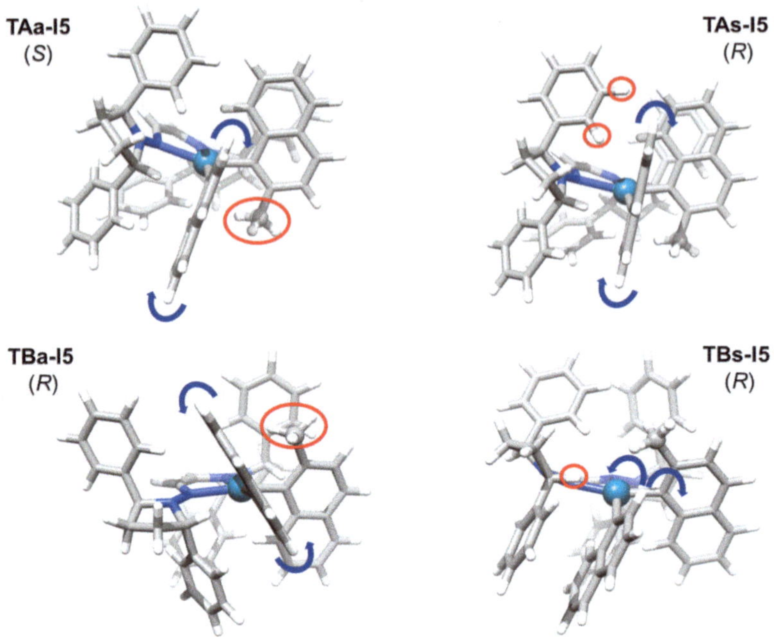

Fig. 6.12 Representation of the steric interactions (in *red*) that govern the stereochemical outcome of the reductive elimination reaction from the intermediates obtained in the four transmetalation routes. Favored rotations of the tail of **3** are shown in *blue*. Only the Me group of **2** is displayed using a ball-and-stick representation for clarity

opposite directions; therefore, the steric bulk of the Ph group of the bis-hydrazone ligand located above the plain defined by Pd and their coordinated ligands is now the responsible for the favored clockwise rotation of **3** and the subsequent formation of the (*R*)-product. Hence, for these three intermediates, the coupling product was the one initially expected (Fig. 6.12).

For the case of **TBs-I5** the results were surprisingly different. In **TBs-I5**, as in **TAs-I5**, the Me group of **2** and the tail of **3** are pointing to opposite directions. However, in this case, the tail of **3** is now below the plain defined by Pd and their ligands, where instead of a Ph group we find a hydrogen atom (due to the C_2-symmetric structure of the bis-hydrazone ligand). Thus, the tail of **3** could, in principle, "easily" rotate both to the right or to the left resulting in both the (*R*)- and (*S*)-products. The estimated energy barrier for the interconversion of the intermediates that give rise to the (*R*)- and (*S*)-products was found to be less than $4\,\text{kcal}\cdot\text{mol}^{-1}$; therefore, the stereochemical outcome of route *B-syn* will depend on the energy barriers for the reductive elimination from those intermediates (Fig. 6.13). Interestingly, the coupling leading to the product with the (*R*)-configuration (**REBs-TS**) was found to have a lower energy barrier, which explains the unexpected configuration for this route

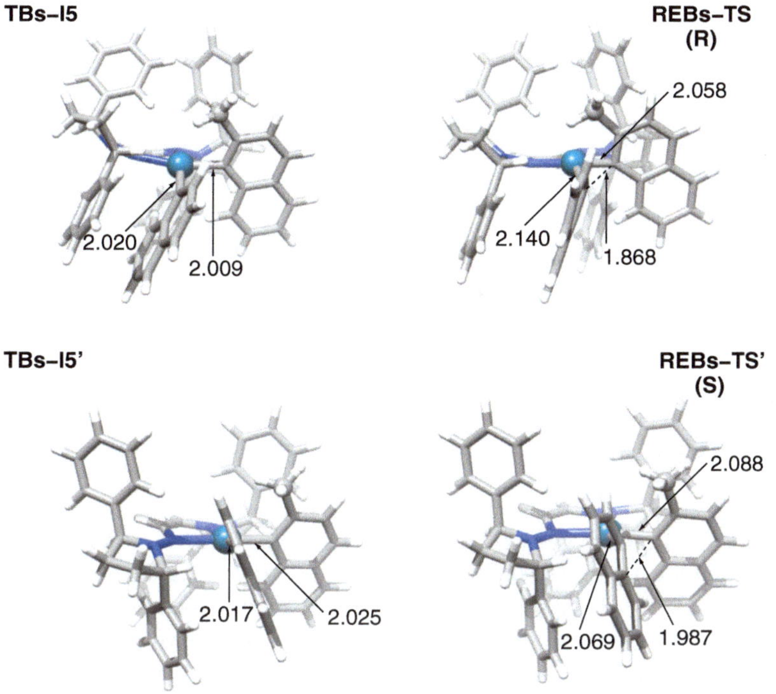

Fig. 6.13 Optimized structures for the intermediates and transition states involved in the reductive elimination in *B-syn* route. Only the Me group of **2** is represented in ball-and-stick for clarity. The stereochemistry of the product that results from each transition is shown in parenthesis

(Table 6.2). The estimation of the energy barriers for the *R*/*S* interconversions for the other routes are underway.

It should be noted, however, that in the case that the barrier for the intermediate interconversions leading to the (*R*)- and (*S*)-products for the other routes were higher than the corresponding energy barriers for the reductive elimination, the enantioselectivity would be already defined in a previous step. But, where it would be exactly defined? In order to find the answer to this question we have to go back to the transmetalation step.

As shown in Fig. 6.14, in the second intermediate of this step (**TBaI2**) the stereochemistry is clearly not defined yet, since **2** and **3** are parallel to each other. Therefore, it necessarily has to be defined in the next step, **TBa-TS2**, where the Pd–C bond with **3** is formed and it can undergo either clockwise or counterclockwise rotations. This transition state, however, is not the one with the highest Gibbs energy barrier, which means that, even if after this step the stereochemical outcome of the route is not modified, this transition state does not govern the absolute configuration observed in the reaction.

In fact, the stereochemistry of the reaction will be mainly dictated by the route with the lowest value for the highest Gibbs energy barrier (**TS1**, Table 6.1). Thus,

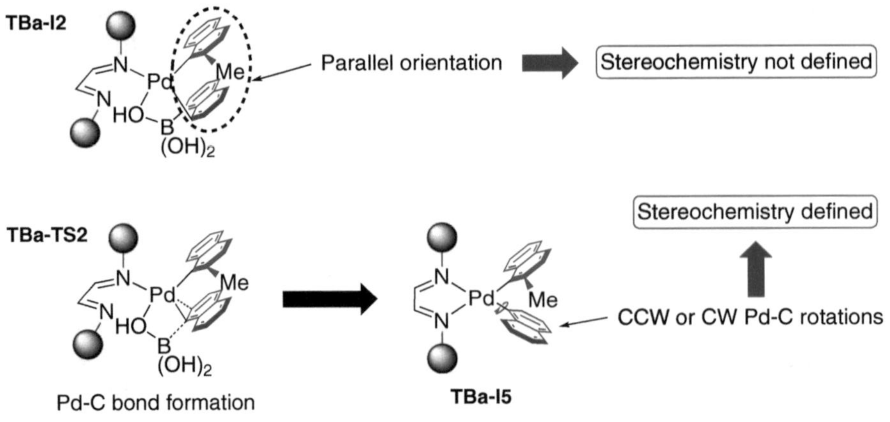

Fig. 6.14 Schematic representation that shows where the stereochemistry is defined

according to this and the values collected in Table 6.1, the (*R*)- or (*S*)-configuration of the final product would be provided by the route *A-anti* (**TAa-TS1**). Consequently, the absolute configuration of the product observed in the experiments should be (*S*). Nevertheless, the product observed in the experiments has a (*R*)-configuration. For this discrepancy between theory and experiments there are two possible reasons: either the true pathway for the reaction has not been identified yet, or before the reductive elimination the interconversion among the intermediates may take place giving rise to the product with the absolute configuration observed in the experiments. In particular, one alternative for the first reason would be that the difference between the rates for the oxidative addition through the routes A and B was enough to prevent the subsequent transmetalation reactions from the former (i.e. *A-anti* and *A-syn* routes). In that case, the route B would be the only accessible route, which would be consistent with experiments as we have previously seen that it only affords the (*R*)-product (Fig. 6.12). On the other hand, an alternative for the second reason would be that the intermediate previous to reductive elimination in the route *A-anti* (**TAa-I5**, Fig. 6.10) evolved to another intermediate that afforded the (*R*)-product with a lower energy barrier. This interconversion would consist in the counterclockwise rotation of the tail of **3** from the left side of the Me group of **2** to the right side of this group. This alternative, in fact, is based on the low energy barrier estimated for the rotation of **3** in the *B-syn* route (Fig. 6.13), and the recently reported study on a related asymmetric Suzuki-Miyaura coupling, where this alternative was deemed responsible for the observed stereochemistry [46].

6.8 Conclusions

In this study, the full catalytic cycle for the recently reported asymmetric Suzuki-Miyaura coupling between 1-bromo-2-methylnaphthalene (**2**) and 1-naphthal eneboronic acid (**3**) catalyzed by a [Pd(bis-hydrazone)] (**1**) complex [47], was theoretically investigated by means of *DFT* calculations. Importantly, the results derived from this study revealed that the transmetalation reaction does not occur in just one step, but occurs in three steps. This is owing to the relative lability of the bis-hydrazone ligand, which can easily dissociate one of the N atoms coordinated to the Pd catalyst. Very recently, this variant for the transmetalation mechanism has been reported for the Suzuki-Miyaura coupling catalyzed by a diimine chelated palladium complex [15].

As suggested in the experimental work [47], four different routes for the reaction leading to the final coupled product were analyzed (i.e. *A-anti*, *A-syn*, *B-anti*, *B-syn*). Furthermore, for each of them, the (*R*)- or (*S*)-configuration of the final product was also determined. Interestingly, theoretical calculations on these routes showed that their stereochemistry is not defined in the step with the highest Gibbs energy barrier (i.e. **TS1**). Moreover, calculations also showed that the (*R*)-configuration observed in the experiments can not be governed by this step, since the route with the lowest Gibbs energy for this step (i.e. *A-anti*) leads to the product with the opposite configuration. The discrepancy between theory and experiments can be attributed to two different reasons: the first one is that the real pathway for the reaction has not been identified yet. One alternative for this reason would be that the rate for the oxidative addition via the route B is higher enough to neglect the oxidative addition through the route A; this would make the route B the only one available for transmetalation, which would give rise to the formation of the (*R*)-product in agreement with experiments. On the other hand, the second explanation can be based on considering that, prior to reductive elimination, the interconversion of the intermediates leading to the (*R*)- and (*S*)-products can easily take place. This type of interconversion has been already observed for the route *B-syn* and also assumed to be responsible for the observed stereochemistry in a related Suzuki-Miyaura coupling [46]. We expect to be able to rationalize the formation of the (*R*)-product observed in the experiments with the calculations we are carrying out.

References

1. Miyaura, N., Yamada, K., Suzuki, A.: Tetrahedron Lett. **20**, 3437 (1979)
2. Miyaura, N., Suzuki, A.: Chem. Rev. **95**, 2457 (1995)
3. de Meijere, A., Diederich, F. In: Metal Catalyzed Cross-Coupling Reactions, 2nd edn. John Wiley & Sons, New York (2004)
4. Tamao, K., Miyaura, N. In: Cross-Coupling Reactions: A Practical Guide, Topics in Current Chemistry, Vol. 219, chap. 1, pp. 1–10. Springer, Berlin (2002)
5. Suzuki, A., Yamamoto, Y.: Chem. Lett. **40**, 894 (2011)

6. Myers, A.G., Tom, N.J., Fraley, M.E., Cohen, S.B., Madar, D.J.: J. Am. Chem. Soc. **119**, 6072 (1997)
7. Suzuki, A.: J. Organomet. Chem. **576**, 147 (1999)
8. Kotha, S., Lahiri, K., Kashinath, D.: Tetrahedron **58**, 9633 (2002)
9. Garg, N.K., Caspi, D.D., Stoltz, B.M.: J. Am. Chem. Soc. **126**, 9552 (2004)
10. Zapf, A., Beller, M.: Top. Catal. **19**, 101 (2002)
11. Suzuki, A.: Angew. Chem. Int. Ed. **50**, 6723 (2011)
12. Braga, A.A.C., Morgon, N.H., Ujaque, G., Maseras, F.: J. Am. Chem. Soc. **127**, 9298 (2005)
13. Braga, A.A.C., Ujaque, G., Maseras, F.: Organometallics **25**, 3647 (2006)
14. Sicre, C., Braga, A.A.C., Maseras, F., Cid, M.M.: Tetrahedron **64**, 7437 (2008)
15. Weng, C.-M., Hong, F.-E.: Dalton Trans. **40**, 6458 (2011)
16. Quasdorf, K.W., Antoft-Finch, A., Liu, P., Silberstein, A.L., Komaromi, A., Blackburn, T., Ramgren, S.D., Houk, K.N., Snieckus, V., Garg, N.K.: J. Am. Chem. Soc. **133**, 6352 (2011)
17. Besora, M., Braga, A.A.C., Ujaque, G., Maseras, F., Lledós, A.: Theor. Chem. Acc. **128**, 639 (2011)
18. Braga, A.A.C., Morgon, N.H., Ujaque, G., Lledós, A., Maseras, F.: J. Organomet. Chem. **691**, 4459 (2006)
19. Carrow, B.P., Hartwig, J.F.: J. Am. Chem. Soc. **133**, 2116 (2011)
20. Ariafard, A., Lin, Z., Fairlamb, I.J.S.: Organometallics **25**, 5788 (2006)
21. Ariafard, A., Yates, B.F.: J. Am. Chem. Soc. **131**, 13981 (2009)
22. Krasovskiy, A., Lipshutz, B.H.: Org. Lett. **13**, 3818 (2011)
23. García-Melchor, M., Ujaque, G., Maseras, F., Lledós, A: In: Peruzzini, M., Gonsalvi, L. (eds.) Phosphorus compounds: advanced tools in catalysis and material sciences, vol. 37, chap. **3**, pp. 57–84. Springer, Berlin (2011)
24. Martín, R., Buchwald, S.L.: Acc. Chem. Res. **41**, 1461 (2008)
25. Christmann, U., Vilar, R.: Angew. Chem. Int. Ed. **44**, 366 (2005)
26. Fleckenstein, C.A., Plenio, H.: Chem. Soc. Rev. **39**, 694 (2010)
27. Goossen, L.J., Koley, D., Hermann, H.L., Thiel, W.: Organometallics **25**, 54 (2006)
28. Joshaghani, M., Faramarzi, E., Rafiee, E., Daryanavard, M., Xiao, J., Baillie, C.: J. Mol. Catal. A: Chem. **259**, 35 (2006)
29. Kozuch, S., Martin, J.M.: ACS Catal. **1**, 246 (2011)
30. Jover, J., Fey, N., Purdie, M., Lloyd-Jones, G.C., Harvey, J.N.: J. Mol. Catal. A **324**, 39 (2010)
31. If you are interested in further reading, a selection of the most relevant computational studies on this topic are discussed in the book chapter (i.e. Article VI) included in Appendix B.
32. Brunel, J.M.: Chem. Rev. **105**, 857 (2005)
33. Berthod, M., Mignani, G., Woodward, G., Lemaire, M.: Chem. Rev. **2005**, 105 (1801)
34. Kozlowski, M.C., Morgan, B.J., Linton, E.C.: Chem. Soc. Rev. **38**, 3193 (2009)
35. Zehm, D., Fudickar, W., Hans, M., Schilde, U., Kelling, A., Linker, T.: Chem. Eur. J. **14**, 11429 (2008)
36. Ma1, B., Zeng1, F., Zheng, F., Wu, S.: Chem. Eur. J. **17**, 14844 (2011)
37. Yin, J., Buchwald, S.L.: J. Am. Chem. Soc. **122**, 12051 (2000)
38. Cammidge, A.N., Crépy, K.V.L.: Chem. Commun. 1723 (2000)
39. Jensen, J.F., Johannsen, M.: Org. Lett. **5**, 3025 (2003)
40. Mikami, K., Miyamoto, T., Hatano, M.: Chem. Commun. 2082 (2004)
41. Colobert, F., Valdivia, V., Choppin, S., Leroux, F.R., Fernández, I., Álvarez, E., Khiar, N.: Org. Lett. **11**, 5130 (2009)
42. Debono, N., Labande, A., Manoury, E., Daran, J.-C., Poli, R.: Organometallics **2010**, 29 (1879)
43. Lee, J.C.H., McDonald, R., Hall, D.G.: Nature Chem. **3**, 894 (2011)
44. Kamei, T., Sato, A.H., Iwasawa, T.: Tetrahedron Lett. **52**, 2638 (2011)
45. Yamamoto, T., Akai, Y., Nagata, Y., Suginome, M.: Angew. Chem. Int. Ed. **50**, 8844 (2011)
46. Shen, X., Jones, G.O., Watson, D.A., Bhayana, B., Buchwald, S.L.: J. Am. Chem. Soc. **132**, 11278 (2010)
47. Bermejo, A., Ros, A., Fernández, R., Lassaletta, J.M.: J. Am. Chem. Soc. **130**, 15798 (2008)
48. Lassaletta, J.M., Alcarazo, M., Fernández, R.: Chem. Commun. 298 (2004)

49. Grasa, G.A., Hillier, A.C., Nolan, S.P.: Org. Lett. **2001**, 3 (1077)
50. Mino, T., Shirae, Y., Sakamoto, M., Fujita, T.: J. Org. Chem. **70**, 2191 (2005)
51. Huang, Y.-L., Weng, C.-M., Hong, F.-E.: Chem. Eur. J. **14**, 4426 (2008)
52. Wang, C., Wu, W.: J. Chem. Educ. **88**, 299 (2011)
53. Zhao, Y., Truhlar, D.G.: Theor. Chem. Acc. **120**, 215 (2008)
54. Frisch, M.J., Trucks, G.W., Schlegel, H.B., Scuseria, G.E., Robb, M.A., Cheeseman, J.R., Scalmani, G., Barone, V., Mennucci, B., Petersson, G.A., Nakatsuji, H., Caricato, M., Li, X., Hratchian, H.P., Izmaylov, A.F., Bloino, J., Zheng, G., Sonnenberg, J.L., Hada, M., Ehara, M., Toyota, K., Fukuda, R., Hasegawa, J., Ishida, M., Nakajima, T., Honda, Y., Kitao, O., Nakai, H., Vreven, T., Montgomery, Jr., J.A., Peralta, J.E., Ogliaro, F., Bearpark, M., Heyd, J.J., Brothers, E., Kudin, K.N., Staroverov, V.N., Kobayashi, R., Normand, J., Raghavachari, K., Rendell, A., Burant, J.C., Iyengar, S.S., Tomasi, J., Cossi, M., Rega, N., Millam, J.M., Klene, M., Knox, J.E., Cross, J.B., Bakken, V., Adamo, C., Jaramillo, J., Gomperts, R., Stratmann, R.E., Yazyev, O., Austin, A.J., Cammi, R., Pomelli, C., Ochterski, J.W., Martin, R.L., Morokuma, K., Zakrzewski, V.G., Voth, G.A., Salvador, P., Dannenberg, J.J., Dapprich, S., Daniels, A.D., Farkas, Ö., Foresman, J.B., Ortiz, J.V., Cioslowski, J., Fox, D.J.: Gaussian 09, Revision A.1, Gaussian, Inc., Wallingford (2009)
55. Fuentes, B., García-Melchor, M., Lledós, A., Maseras, F., Casares, J.A., Ujaque, G., Espinet, P.: Chem. Eur. J. **16**, 8596 (2010)
56. García-Melchor, M., Fuentes, B., Lledós, A., Casares, J.A., Ujaque, G., Espinet, P.: J. Am. Chem. Soc. **133**, 13519 (2011)
57. Andrae, D., Häussermann, U., Dolg, M., Stoll, H., Preuss, H.: Theor. Chim. Acta. **77**, 123 (1990)
58. Ehlers, A.W., Bohme, M., Dapprich, S., Gobbi, A., Hollwarth, A., Jonas, V., Kohler, K.F., Stegmann, R., Veldkamp, A., Frenking, G.: Chem. Phys. Lett. **208**, 111 (1993)
59. Check, C.E., Faust, T.O., Bailey, J.M., Wright, B.J., Gilbert, T.M., Sunderlin, L.S.: J. Phys. Chem. A **105**, 8111 (2001)
60. Marenich, A.V., Cramer, C.J., Truhlar, D.G.: J. Phys. Chem. B **113**, 6378 (2009)

Chapter 7
General Conclusions

The whole of science is nothing more than a refinement of everyday thinking

Albert Einstein

The specific conclusions for each of the cross-coupling reactions investigated in this thesis have been presented at the end of the corresponding chapter. Thus, only the general conclusions drawn from these studies will be summarized herein.

In the case of the Negishi reaction, the theoretical studies on the transmetalation reactions with ZnMeCl and $ZnMe_2$ showed the complexity of these processes and provided a detailed picture of their reaction mechanisms. In particular, for the transmetalation reaction with ZnMeCl, calculations warned on the many chances for the generation of new intermediates that would eventually afford homocoupling side products. On the other hand, the theoretical results derived from the study with $ZnMe_2$ proved the existence, operation and effects of competitive transmetalation pathways, some of which had not been invoked before, but that should be considered when planning or discussing Negishi syntheses.

For the copper-free Sonogashira reaction, the mechanistic study reported in this thesis revealed that, just like in other cross-coupling reactions (i.e. Stille, Negishi), there are several competing reaction pathways and a change on the reaction conditions (e.g. solvent, ligands, substrates, base) might favor one over the other ones. Moreover, a new mechanism in which the acetylide (formed by deprotonation of the alkyne) directly reacts with the catalyst was also proposed.

Last but not least, theoretical calculations on the investigated asymmetric Suzuki-Miyaura coupling showed that the mechanism for the transmetalation step differs from the typical reaction pathway proposed in the literature. This is owing to the relative lability of the bis-hydrazone ligand, which can easy dissociate one of the N atoms directly coordinated to Pd. As far as the stereochemistry of the reaction is concerned, calculations so far do not provide an explanation for the high enantioselectivities observed in the experiments. Two possible reasons are: either the true pathway for the reaction has not been identified yet, or an interconversion among

M. G. Melchor, *A Theoretical Study of Pd-Catalyzed C–C Cross-Coupling Reactions*, Springer Theses, DOI: 10.1007/978-3-319-01490-6_7,

the reaction intermediates takes place prior to the reductive elimination giving rise to the product observed experimentally.

Overall, theoretical calculations have been used in this thesis to determine, elucidate, and propose reaction mechanisms for Pd-catalyzed cross-coupling reactions. In particular, they have allowed the characterization of reaction intermediates and transition states involved in these processes. Hence, we can conclude that the results presented in this thesis prove that the use of computational methods, namely *DFT* calculations, is a very useful tool for the study of reaction mechanisms of homogeneous catalytic reactions.

MIX
Papier aus verantwortungsvollen Quellen
Paper from responsible sources
FSC® C105338

If you have any concerns about our products,
you can contact us on
ProductSafety@springernature.com

In case Publisher is established outside the EU,
the EU authorized representative is:
Springer Nature Customer Service Center GmbH
Europaplatz 3, 69115 Heidelberg, Germany

Printed by Libri Plureos GmbH
in Hamburg, Germany